Rereading the Machine in the Garden

North American Studies

Edited by the The John F. Kennedy Institute for North American Studies of Freie Universität Berlin, the Center for North American Studies (ZENAF) of Goethe-Universität Frankfurt am Main and the Department of English and American Studies of Humboldt-Universität zu Berlin.

Volume 34

Eric C. Erbacher is a Lecturer in American Studies at the University of Muenster. *Nicole Maruo-Schröder* is Professor of Cultural Studies at the University of Koblenz-Landau. *Florian Sedlmeier* is Assistant Professor of American Literature at the John F. Kennedy Institute for North American Studies, Freie Universität Berlin.

Eric Erbacher, Nicole Maruo-Schröder,
Florian Sedlmeier (eds.)

Rereading the Machine in the Garden

Nature and Technology in American Culture

Campus Verlag
Frankfurt/New York

Bibliographic Information published by the Deutsche Nationalbibliothek.
The Deutsche Nationalbibliothek lists this publication in the Deutsche Nationalbibliografie; detailed bibliographic data are available in the Internet at http://dnb.d-nb.de
ISBN 978-3-593-50191-8

Cover design: Campus Verlag, Frankfurt-on-Main
Cover illustration: © Olafur Eliasson, "Viewing Machine", photo by Alexandre S. Dias, 2013 photo courtesy of Alexandre S. Dias.
Typeset: Campus Verlag, Frankfurt-on-Main
Printed on acid free paper.
Printed in the United States of America

This book is also available as an E-Book.
www.campus.de
www.press.uchicago.edu

Contents

Introduction: Rereading *The Machine in the Garden*

Eric Erbacher, Nicole Maruo-Schröder, Florian Sedlmeier

Marking an anniversary, such as the fiftieth of the publication of Leo Marx's *The Machine in the Garden: Technology and the Pastoral Ideal in America* (1964), one creates and singles out an event, suggesting a need to return to this event and to reassess its implications for past, present, and future conditions. In this sense, the notion of an anniversary carries the connotations of both a celebration and a re-envisioning. In and of itself, however, the marking of an anniversary does not necessitate a critical reexamination. After all, one could leave the respectful tone of the anniversary and easily dismiss Marx's study as a relic of bygone times, guilty of a range of inadequacies that subsequent critics have exposed in a plethora of articles and books. Similar to other representatives of what Bruce Kuklick (1972) labeled the "Myth and Symbol School," Marx has been charged with participating in the tale of American exceptionalism that continues to haunt American Studies, a tale grounded in the sermons of John Winthrop, the political rhetoric of Thomas Jefferson, and others; with rearticulating the belief in a holistic national ideal; with glossing over discourses of difference and with not systematically accounting for their effects on the societal fabric; and with foreclosing the possibilities of the recently fashionable and reactivated transnational approaches. These political objections are sometimes complemented by a critique of the philosophical underpinnings of Marx's argument, most notably of what Kuklick identified as its pervasive humanism and its supposed indebtedness to Cartesian thought, i.e. its marked separation of thought and the material.[1] And they find their aesthetic equivalent in the critique of an understanding of the arts and literature that is regarded as suspicious because it positions the declared masterpieces of the American Renaissance and modernism as privileged objects

1 While Kuklick's article became a point of reference for revisionist scholars if mostly because it coined the label, Alan Trachtenberg (1984) has provided perhaps the most succinct defense of the Myth and Symbol School.

of study and because it conceives of these texts as being structured according to a single master trope: the machine in the garden. In order to substantiate his claim, critics have argued, Marx not simply ignores a large number of texts (both popular and 'high literary' ones); John Lark Bryant criticizes him for deploying "extravagant argumentation" (1975, 68) in his readings of Renaissance masterpieces to position the machine in the garden as the one tropology that best describes America.[2] The validity of many of these arguments depends, of course, on the degree and mode of historicizing them, i.e. on examining them for the ideologies they necessarily contain. And if we seek to legitimize this volume by marking not just *The Machine in the Garden*'s anniversary but also its continuing significance, we inevitably engage in modes of historicizing as well.

The well-known charges just sketched are indicative of the narrative of a generational shift from Americanists to New Americanists, a term coined by Frederick Crews.[3] This shift manifests itself, among others, in a changing understanding of the political function of scholarship, in a different conception of scholarly argumentation and writing, and in the emergence of theory.[4] Against this backdrop, the present collection of essays reexamines *The Machine in the Garden* on two basic grounds. First, the volume, and this introductory essay in particular, aim at historicizing the study by redirecting the well-known and necessary critiques of the New Americanists, informed by cultural studies, gender studies, new historicism, ethnic studies, and poststructuralist theory. Such a task of historicizing is not least suggested by Marx himself, who, challenged by other scholars, reevaluated his own and his generation's positions in the changing field of American Studies, defending, expanding, and refining his initial argument in a series of articles, some of which are included in his collection of essays *The Pilot and the Passenger* (1988).[5] At stake in this project are shifting codifications of

2 See also the reviews of Steiner (1964) and Mills (1970) for a criticism of Marx's method of *literary* analysis as well as the scope of analysis and selection of texts.

3 See Pease (1990, 1); see also e.g., Lipsitz (2001), Rowe (2002) for discussions of this paradigm shift.

4 For an up-to-date discussion of the generational shift from the integrationist Americanists to the revisionist, pluralist New Americanists see Johannes Voelz (2010), who shows that this shift relies on the narrative, or "the thesis of an epochal break" (23), which he finds reiterated in accounts by Gene Wise (1979), Donald Pease and Robyn Wiegman (2002), and Leo Marx (2005), among others. For a discussion of the function of theory in this narrative of a shift see also Voelz (2010, 24–27).

5 Later writings include Marx's "Afterword: The Machine in the Garden" (1999/2000), included in the thirty-fifth anniversary edition; several articles on technology (1984;

the arts and literature as well as culture, nature, and technology. Second, the contributions to this collection seek to retrieve the trope of the machine and its intrusion into the pastoral landscape as a vital configuration for a broad range of artistic, filmic, and literary texts, spanning from the early nineteenth to the early twenty-first centuries. While the trope, for various reasons, can certainly not account for the grand cultural and literary history Leo Marx had in mind, its significance in artistic productions from a U.S. context has to be acknowledged and construed. The present volume probes the unabated relevance of this cultural tropology for the analysis of representations of nature and technology in artistic, filmic, and literary texts. And in doing so, this introduction and the contributions also assess the lasting impact of Marx's method and rhetoric for the present condition of the field of inquiry named American Studies, looking for points of entry that *The Machine in the Garden* might retrospectively offer for current debates.

Institutional Concerns: American Studies as a Field of Inquiry

In a sense, the poles of Marx's dialectical trope are inscribed in three institutions: Marx is a Harvard graduate, who received his PhD in History of American Civilization in 1950, and he held academic positions at the University of Minnesota—hence in the upper part of the Midwest, that geographical and cultural region most commonly associated with notions of the middle landscape and the heartland—and at the Massachusetts Institute of Technology in Cambridge, where he was appointed Kenan Professor of American Cultural History in 1976. The denominations and trajectories of these institutional appointments should make us aware of a key aspect of the project of American Studies: it was initially conceived as a field of inquiry geared toward identifying the possibilities and the structures of a cultural history. As such, American Studies, in its inception, is not a discipline but a specific method; it is positioned against the empiricism and scientific rigor of the natural and social sciences, and it is devoid of the philological tradition that continues to inform European languages

1997); an essay that contextualizes the early Americanists in the institutional context of the University of Minnesota (1999); and an exchange across generations of Americanists with George Lipsitz in *American Literary History* (see Lipsitz 2005; Marx 2005).

and literatures and its academic institutions. In effect, Marx, his teacher Henry Nash Smith and others repeatedly insisted on this peculiarity of the emerging field, not the least in order to legitimize its specific position in the departmental organization.

Smith poses the central institutional and methodological question that concerned early Americanists in the title of his influential essay: "Can American Studies Develop a Method?" (1957). Reading Mark Twain for his various prose styles and the social positions they represent, Smith calls for "a method of analysis that is at once literary [...] and sociological" (201). Partly drawing on Matthew Arnold, he positions the conceptual metaphor of culture "to embrace 'society' and 'art'" (ibid., 206). Culture is presented as a third alternative that avoids the pitfalls of the social sciences, which hold that "all value is implicit in social experience, in group behavior, in institutions, in man as an average member of society," and which questions the assumption of New Criticism "that value lies outside society, in works of art which exist on a plane remote from [...] our actual experience" (ibid., 206). The construction of culture as a third conceptual metaphor between these notions of the social sciences and 'purely' literary criticism allows Smith "to conceive of American Studies as a collaboration among men working from within existing academic disciplines but attempting to widen the boundaries imposed by conventional methods of inquiry" (ibid., 207). Anchoring the field in "practice" rather than method, he means "to resolve the dilemma posed by the dualism which separates social facts from esthetic values" and suggests to explore the mental and symbolic structures, or "the image in our minds" (ibid., 207) in which both supposedly merge; the singular, of course, betrays the search for and the belief in a single, unifying "image." American Studies, for Smith, emerge from within the institutionalized disciplines, breach their methodologies, and put their lines of demarcation to a practical test. As such, the field is per definition envisioned as inter-disciplinary in the perhaps truest sense of the word: it is between the disciplines, from within the institutional confines of which it intervenes, without being and becoming a discipline of its own.

Smith's inquiry reverberates in Marx's writings. In effect, most of the articles he published after *The Machine in the Garden* are concerned with institutional and methodological questions, starting with "American Studies: A Defense of an Unscientific Method" (1969). In "Reflections on American Studies, Minnesota, and the 1950s" from 1999, he reviews the project of his generation vis-à-vis both the institutionalized New Criticism and the shift toward cultural studies. He locates the beginnings of American Studies "outside the academy in the work of independent, unaffiliated writers and intellectuals" (44), among whom he counts thinkers as diverse as W.E.B. Du Bois and Lewis Mumford. Explaining the subsequent institutional divide in the English department at Minnesota between the contextualism and historicism practiced by Americanists and the closed textualism of the New Critics as well as the reactionary parochialism of the Southern Agrarians, he argues that the actual impact of the divide was smoothened by the identification "of a common enemy: McCarthyism and other hyper-nationalistic expressions of paranoid anti-communism" (ibid., 42). The stakes in the emergence and institutionalization of American Studies were not least the values of intellectual independence in a political climate of censorship. And these values manifested themselves in the search for a meaningful structure of myths and symbols that might reclaim a national cohesion as a cultural one from the governmental and medial hijacking of the nation in paranoid nationalistic terms. The precariousness of this constellation accounts for a positioning such as "secular, left-liberal humanist values" that Leo Marx ascribes to Americanists of his generation, whom he deems "committed to the labor movement, to 'progressive' [...] principles of social and economic justice" (ibid., 41). Against the backdrop of the 1950s, which betrayed the political legacy of the New Deal, Marx reclaims "the nation's distinctiveness" as grounded in "singular political innovations" (ibid., 43). Note the choice of vocabulary: it is about distinctiveness and singularity, not about exceptionality and uniqueness. Elaborating upon Lincoln's Gettysburg Address, he further contends: "the United States is a nation defined neither by its location nor its ethnic composition, but rather by a 'proposition'—a cosmopolitan, multicultural, potentially universalizable set of principles" (ibid., 43).

This intricate piece of rhetoric testifies to the conceptual struggles Marx faces in reexamining and retroactively legitimizing his project. The United States expresses an ideal and a promise, tied neither to a definition of the nation as based upon the geography of the nation-state, nor to an understanding of the nation that would be contingent upon ethnic difference. Rather, the idea of a nation that Marx has in mind aspires both to a notion of intellectual cosmopolitanism, which might effectively recode the nation as a non-nationalist organism composed of a single community of citizens of the world, and to an understanding of multiculturalism that respects cultural variety without prioritizing ethnic difference as a category of analysis. This implies not to give up on aspirations to liberal consensus in the framework of a reactionary narrative of 1950s Cold War consensus: Marx reclaims a leftist liberalism from within and against the liberalisms of libertarian economics, consumer culture, and limited government. Consequently, he stresses two institutional and two societal motivations for the project of American Studies. Inside the university, Americanists strove for "the introduction of distinctively American subject matter in the study of history, literature, philosophy, and art" (ibid., 45) in order to defy "snobbish Anglophilia" (ibid., 46); this challenge to the elitist literary standards of English departments resulted in "a desire to circumvent barriers to interdisciplinarity" (ibid., 46). Outside departmental reform, Marx goes on to argue, Americanists were "essentially ideological" (ibid., 46) to the extent that they committed themselves to the complex retrieval of the foundational ideas just delineated.[6] In doing so, they sought to apply "democratic standards of multicultural equity in recruiting faculty members," which during the 1950s meant an extension of the pool of WASP males "to candidates of Jewish, Irish, German, Polish, and other non-WASP white males" (ibid., 48).

For Marx, "the affirmative commitment" that he deems the unifying characteristic of his generation of scholars faces a challenge, a "crisis of legitimacy that developed in 1968–75" (ibid., 50, 49). This crisis fosters a renewed skepticism toward the idea of national-cultural holism—a skepticism manifesting itself in what Marx calls an "activist political radicalism of the academy," which corresponds to a "conceptual radicalism characteristic of post-structural critical theory, social constructivism, and their variants" (ibid., 49). For him, the "affirmative commitment" now shifts from presumably macroscopic to supposedly microscopic scales of analysis. We could certainly criticize such a view for reproducing an anxiety of the fractions and frictions that characterizes the so-called "culture wars" of the 1980s and 1990s, and we could equally contend that the developments which grew out of the new social movements are no less and maybe even more macroscopic. But we can also take Marx's argument as a point of

6 The retrospective concession of being openly "ideological" reacts to the various charges of ideological complicity New Americanists directed at Marx. His embrace here is not without irony once we keep in mind that declaring oneself to be an ideologue of sorts—and hence admitting the inevitability of ideological complicity and cooptation—marks a scholarly taboo in the context of American Studies in the U.S.

entry for reexamining the politics of experience that is by now firmly institutionalized. After all, the institutionalization of discourses of difference and identity politics as paradigmatic preconditions for the production of certain literary histories often commands a strategic employment of the memoir, which is used as a mode of writing and reading to validate arguments and their representativeness for communities.[7] The lesson to be learned here, from a contemporary point of view, rather concerns the persistency of "affirmative commitment" as a strategy in scholarship across the shifts of scope.

The Machine in the Garden: A Cultural Tropology and its Intellectual Sources

The Pastoral

One way to historicize *The Machine in the Garden* more thoroughly consists in reconstructing the key concepts, the intellectual sources and their relation to Marx's research design. His interest in the technological and the pastoral can be explained by looking at the institutionalized intellectual traditions of his contemporaries. When Marx resorts to the pastoral as both a conceptual metaphor and a literary convention, he inscribes himself into an institutionalized discourse of English departments, where scholars such as William Empson and Frank Kermode established this category of analysis from the 1930s through the 1950s, and which Raymond Williams would take up again in the 1970s. Leo Marx, then, operates from within and against a master trope that goes back to antiquity and that figures prominently in the debates of English New Criticism, which was notably different from its American variant in its thinking of the relation between art, politics, and society.

In *Some Versions of the Pastoral* (1935), William Empson develops an understanding of proletarian literature as pastoral in the sense of "folk-litera-

7 Werner Sollors's work is a vital reminder that the writing of literary histories according to discourses of difference and identity politics, which we now almost take for granted, is but one way of writing such histories (see 1986, 14); among others, Kenneth Warren underscores the central significance of the memoir for legitimizing and administering African American studies as a field of inquiry (see 2011, 122).

ture," written "by the people, for the people, and about the people;" for Empson, such a broad conception of proletarian literature is "not dependent on a system of class exploitation" (1966, 15). "The essential trick of the old pastoral," he writes, was to bring forth "a beautiful relation between the rich and the poor" (ibid., 19), based upon representations of feelings that were "common to all classes" (ibid., 18). As such, it harbors a unifying, reconciliatory social quality that, in Leo Marx's reading of Empson in his chapter on the garden, "served as a mask of political purposes" (2000, 129). Marx adds an asterisk that features, against Empson, his own distinction "between those versions of pastoral which enable the reader to enjoy an easy resolution of the conflict and those which enforce the poet's ironic distance from the pastoral dream, that is, between sentimental and complex pastoralism" (ibid., 129). This argument is framed by a reading of "Jefferson's idealized portrait of the husbandman," whom Marx takes to be a cultural translation of "the literary shepherd" (ibid., 130). Even more: the credibility of "this democratic Everyman," he writes, derives from the circumstance that "the pastoral ideal has been so well assimilated into an American ideology" (ibid., 130). Then comes the moment where Marx makes clear the specificity of the pastoral for American culture vis-à-vis Empson's reading of the pastoral as an allegory of reconciling economic class differences. He positions the democratic everyman as one who

> achieves the political results outside literature formerly achieved by the shepherd. In the age of [Andrew] Jackson there no longer will be any need to insinuate a beautiful relation between the rich and the poor. By his mere presence the "common man" threatens—or promises—to supplant them both. In the egalitarian social climate of America the pastoral ideal, instead of being contained by the literary design, spills over into thinking about real life. (ibid., 130)

Two brief observations—one conceptual, the other methodological—are important. On the conceptual side, the difference Marx draws between the continental pastoral tradition and the American version echoes the familiar tale of America as a classless society. It is decisive that Marx writes about an "egalitarian *social* climate." This explains the ambiguities at stake in the quoted passage. The figure of the common, agrarian, and egalitarian man is socially classless because it does away with factors such as family name, prestige, etc. And as such, this figure holds the promise to be socially classless from its inception. At the same time, it "spills over" into "real life" and threatens to disguise the nevertheless existing socioeconomic class differences instead of reconciling them. Both the threat and the promise of Thomas Jefferson's poeticized political rhetoric show in the aspired actualization of a transposed and modified literary figure as a political and societal vision. On the methodological side, Marx's literary reading of Jefferson's political rhetoric serves to draw a line of demarcation to the literary analyses of Empson. It once again positions American Studies as being interested in a broad spectrum of texts, downplaying the institutionalized importance of canonized poetry and extending the methods of literary analysis to cultural phenomena.

The other scholar of the pastoral that can most explicitly be linked to Leo Marx is Frank Kermode. In *English Pastoral Poetry* (1952), Kermode reads English Renaissance poetry and positions the pastoral, conceived as the simple, the primitive and the natural, in opposition to the cultivated. And while he concedes that "this opposition can be complex," he argues that "the bulk of pastoral poetry" rests upon the assumption that "natural men are purer and less vicious than cultivated men" (1952, 19). When Kermode writes that "Pastoral is one of the 'kinds' of poetry, like Epic, Tragedy, and Satire" (ibid., 11), he conceives of it as a mode of representation that foregrounds the relationship between art and nature; and the shift that the Renaissance introduces vis-à-vis antiquity is an understanding that "Nature had been enlarged by new knowledge" insofar as pastoral poetry no longer focused on "the happy peasant or shepherd, but the true natural man of the New World" (ibid., 43).

Kermode's 1954 introduction to The Arden Edition of William Shakespeare's *The Tempest* (1611, 1623), a play informed by Montaigne's essay "Of Cannibals" (1580), establishes the link between the pastoral and colonial history more clearly—if not systematically enough for later scholars such as the new historicist Stephen Greenblatt (1992) or the postcolonial critic Ania Loomba (1989; 2002), who works out the implications of racial discourse neglected in the accounts of Kermode and others. Kermode classifies *The Tempest* as a "pastoral drama [...] concerned with the opposition of Nature and Art" mapped onto Caliban and Prospero, respectively (1954, xxiv). The classification of the play as pastoral leads Kermode to contend: "Caliban is the ground of the play" because he functions "to illuminate by contrast the world of art, nurture, civility" (ibid., xxv). With the foregrounding of Caliban the concept of nature is transformed because he is not "a virtuous shepherd" but "a salvage and deformed slave" (ibid., xxxviii). Contending that Montaigne and Shakespeare develop an economy of representations that "describe the native as purely virtuous or purely vicious," Kermode brings forth his central argument that the pastoral is mapped onto the primitive in the literatures that explore the conquest of the New World: "Literary men," he writes, "saw in the favourable reports a rich affirmation of a traditional theme of poetry" (ibid., xxxvi). The recoding of the pastoral as the primitive is enacted in the figure of Caliban, who becomes "an inverted pastoral hero, against whom civility and the Art which improves Nature may be measured"—a measurement that oscillates between an affirmation of the superiority of art, understood as a mechanism of disciplining emotions and producing knowledge, and a demonstration of its corruption (ibid., xlii). In any case, with the Renaissance and the colonization of the New World the pastoral, for Kermode, becomes an adaptable convention of representation that traverses genres—travelogue, play, poetry—and cultural contexts.

Leo Marx merely mentions Kermode's analysis in a footnote without discussing it, when he opens his own reading of *The Tempest* in the second chapter of *The Machine in the Garden*.[8] He conceives Shakespeare's play "as a prologue to American literature" (Marx 2000, 72). The inclusion of *The Tempest* serves several purposes at once. First, institutionally, it anchors the project of American Studies in the conventions of the English department. Marx even rewrites parts of English literary history as American. At the same time, in light of debates about the beginnings of American literary history, the inclusion of Shakespeare as a prologue to that history establishes Marx as a precursor to scholars such as William Spengemann (1984), who positions Aphra Behn's *Oroonoko* (1688) as the first American novel, arguing against the writing of a literary history that rests upon the idea of the nation-state. While he retrieves and returns to the ideals of Jefferson, Marx's interest is less in a national literature than in a structure of tropes that might enable the project of a national-cultural literary history, contingent upon a transcultural validity and a cultural transferability of these tropes. Second, if Marx positions *The Tempest* as a prologue to American literary history—belonging to that history by means of a foreshadowing—this literary history is inevitably inscribed into colonialism. To be sure, Leo Marx can neither perform a poststructuralist nor a postcolonial critique, at least not in the institutionalized version we know today. Like Kermode, he rather focuses on the transposition of a literary convention, albeit with some significant modifications.

8 The first chapter unfolds from the reading of a notebook entry by Nathaniel Hawthorne, which James Dorson takes up in his contribution to this volume.

Key to Marx's claim that *The Tempest* prefigures American literature is his contention that during "the age of discovery [...] a note of topographical realism entered pastoral;" at the time of the English Renaissance, he writes in line with Kermode, authors of travel literature began "to set the action in a terrain that resembled, if not a real place, then the wish-colored image of a real place" (ibid. 47). The "note of topographical realism," then, emerges from the setting of the action in "an actual New World" (ibid. 47), and it facilitates what Marx calls a "genetic" connection between Shakespeare's play and American literature. For him, the "topography of *The Tempest* anticipates the moral geography of the American imagination" (ibid. 72).

How does Marx conceive this "genetic" link between colonial topographic realism and American moral geography? If Frank Kermode looks at the play largely in terms of a Manichean opposition between Caliban and Prospero, Marx complements the Caliban-Prospero constellation more elaborately with a differentiation between Gonzalo and Prospero. In contrast to Kermode, he insists upon a "difference between a pastoral and a primitive ideal" (ibid., 65). While Gonzalo projects the utopian vision of a primitive society, Prospero "stands on a middle ground, a terrain of mediation between nature and art, feeling and intellect" (ibid., 65). The complex pastoral he represents emerges from the contradiction between the notions of art, i.e. his position as "reclusive scholar" turned "social engineer" (ibid. 56), and nature, here associated with a structure of emotion that informs his and his court's retreat from the city of Milan to the remote island. If Kermode establishes a dichotomy between Caliban and Prospero, Marx identifies the relation between art and nature as an ambiguous dialectic, holding a vague promise of reconciling the poles by renouncing both in their pure form. He thus complements his "genetic connection between *The Tempest* and America" with one of prophecy or promise that relates to the notion of an ambiguous moral geography (ibid., 68). The decisive aspect for him remains the regenerative, ritualistic and "redemptive journey away from society in the direction of nature," which "offers the chance of a temporary return to first things" (ibid., 69). This colonial structure, Marx contends, resurfaces in texts such as *Walden*, *Moby-Dick*, and *Huckleberry Finn*. And this claim ultimately allows him to construct his literary genealogy.

The temporariness of the retreat to nature that characterizes the plot of these three texts is significant because it disables both a permanent longing for the primitive and an untainted return to the corruption of the city, which is representative of various technologies. Marx links the "pastoral hope" that emerges from this temporariness to the promise of "political reform," which consists in the "symbolic middle landscape created by mediation between art and nature" (ibid., 71). Such an understanding of political reform leads him back to "Jefferson's vision of an ideal Virginia, an imaginary land free both of European oppression and frontier savagery" (bid., 72). Of course, in the final analysis, this vision, whether Prospero's or Jefferson's, requires an erasure of the cultural 'other' by means of an economy of colonial representation that remains Manichean insofar as the cultural 'other' functions as a structure of fantasy. But Marx's interest lies elsewhere, and his assessment of the possibility of actualizing a political reform that might reconcile art and nature—the liberal vision of a middle ground—is skeptical, to say the least. While he identifies the structural analogy of the ritualistic journey in Shakespeare's play and "our typical American fables," he writes that the "American hero [...] often is further than Prospero from envisaging an appropriate landscape of reconciliation" (ibid., 72). While the prophetic vision or promise produces a powerful and elaborate cultural tropology, and while American writers in Leo Marx's reading return and cater to this promise and its tropology, its actualization remains a fragile affair, fraught with contradictions.

If we bring his reading back to the historical context of *The Machine in the Garden*—the erosion of the New Deal, McCarthyism, and the Cold War consensus narrative—we can identify both his insistence on maintaining the promise of a liberal middle ground and his skepticism that such a promise of procedural political reformism can be realized. Beyond that, the major historical achievement of Marx's exploration of the pastoral vis-à-vis Empson, Kermode and other English literary historians lies in the dialectical conceptualization of the tension between notions of nature and an understanding of the technological that includes the arts, industrial capitalism, and science. The ensuing section aims at assessing these multiple codifications of technology in Marx's writing.

The Technological

The contradiction and attempted reconciliation of art and nature that Marx identifies in *The Tempest* prepares his readings of the garden (mostly in

Thomas Jefferson and Héctor St. John de Crèvecœur) and the machine in the ensuing chapters. The section on the machine moves from Jefferson, who conceives of the machine as "a token of that liberation of the human spirit to be realized by the young American Republic" (ibid., 150), to the merchant Tench Coxe, for whom "the machine is the instrument and not in itself the true source of America's future power" (ibid., 155), to Thomas Carlyle and Timothy Walker onto a web of readings of artistic, literary, popular and philosophical texts by artists and writers such as George Inness, Alexis de Tocqueville, and Walt Whitman.

In order to better understand Leo Marx's broad notion of the technological, a look at his reading of Carlyle's "Signs of the Times" (1829) and Walker's reply "Defence of Mechanical Philosophy" (1831) is insightful, not least because his construal situates the discussion about the valences of technology in the context of different national-cultural codifications that lead back to the institutionalization of American Studies. The opposition between Carlyle and Walker implicitly sets up a difference between English and American cultural imaginations. Carlyle proclaims "the Mechanical Age" (1829, 442). His central complaint is that "the mechanical genius of our time" does not merely affect "the external and physical" but also "the internal and spiritual" domains, pervading the branches of knowledge from science to art and in effect the whole political and social organization, hence undermining the notion of humanism (ibid., 442). Marx not only clearly carves out that Carlyle's intervention aims at the doctrines of eighteenth- and nineteenth-century utilitarianism and environmentalism, he also identifies a dual conception of "the machine as object (a technological fact) and the machine as metaphor (a token of value)" in Carlyle, whose approach he likens to that of "a modern anthropologist" (2000, 173). He thus inscribes himself into this holistic cultural criticism and the use of the machine as a symbolic structure, legitimizing his own project once again through and against English intellectual history. He creates a point of departure from Carlyle, however, when he takes his complaint about "the subordination of the 'dynamical' to the 'mechanical' aspects of life" as anticipating "the post-Freudian version of alienation" which would provide the adequate vocabulary of both culture and self that Carlyle's early-nineteenth-century morals still lack (ibid., 178).

It is the continuity of the narrative of cultural and individual alienation qua mechanization that Marx identifies and juxtaposes with Timothy Walker's American reply. Walker rejects Carlyle's anxiety of cultural decline; he celebrates and embraces the machine as epitomizing human progress, elevating it to the status of a promise and a prophecy. Marx stresses that the valence of the machine, for Walker, is political insofar as he conceives "the new technology as the instrument appointed to fulfill the egalitarian aims of the American people" (ibid., 187). In this context, he writes about the Industrial Revolution—an "irrepressible epithet" at the beginning of the nineteenth century, which "expresses the close kinship between two new forces, political and technological" (ibid., 187). In a footnote marked with an asterisk, he explains his own position and the use of this label, making clear that the question of cultural historians differs from that of economic historians. While the latter inquire "whether there was a break [...] so severe as to justify the name 'revolution'," cultural historians like himself "recognize that we are dealing with a metaphor," the appeal of which can be found in its omnipresent "vivid suggestiveness" (ibid., 187). Conceiving of the machine and the Industrial Revolution as metaphors, Marx is able to view them as "a property of the general culture," as a cultural imaginary so to speak (ibid., 191). Before criticizing Leo Marx for repeating and reinforcing the mythology he proclaims to dismantle, we should bear in mind that his project is above all an analysis of cultural tropes against the backdrop of the empiricism of the social sciences. And it may be a vital reminder today to appreciate the core competencies of cultural and literary criticism amidst the current trend to cater to the rhetoric of innovation and embrace the metaphors and methodologies of the social sciences in the race for funding.

Setting aside these aspects, we can say that Marx mobilizes the Carlyle-Walker constellation in order to distance himself from both approaches to the technological. For him it is not a decision whether the technological is a progressive and prophetic promise or an alienating force, whether it is inclined to liberate or to control. Rather, the task of the cultural historian is first and foremost to expose these diverging valences and unfold their complexity, which emerges dialectically from technology's interplay with the pastoral but also from its semantic overflow that, in turn, might deplete the concept.

During the 1980s, the decade when he was hired by the Massachusetts Institute of Technology, Leo Marx tried to clarify and substantiate his notion of the technological as a semantic field of sorts, which includes the capitalist mode of production, the methods of the empirical social sciences, and the arts. Engaging Martin Heidegger and his essay "The Question

Concerning Technology" (1954), for whom technology manifests a cultural practice and tendency of enframing and revealing (1977, 5–6, 9–10), Marx cites from the etymological analysis Heidegger provides. The notion of *techne*—encompassing craft and skills, the arts of the mind, but also the fine arts—produces a semantic complexity that serves Marx to legitimize his own understanding of the technological as highly ambiguous and ambivalent; the proximity of *techne*, *episteme* and *poesis* in Greek thought, which are "closely related, if not interchangeable, names for knowing in the widest sense" adds to this impression (Marx 1984, 641). These semantic conjunctions and correlations, which foreground the production of knowledge, allow him to engage a form of cultural criticism that once again is attentive to circulating metaphors and their codifications.

Marx discusses Heidegger's cultural diagnosis that modern technology is contingent upon enframing, a "rationalistic and instrumental cast of mind, or form of social practice" (ibid., 643). He sides with Heidegger's main argument that the essence of technology is not technological. And because this is the case there is no inherent danger in technology. The problem resides in the social context and the utilization of the technological. Working out the details in Heidegger's complex understanding of technology and Marx's reading and misreading of Heidegger would warrant another article. Suffice it to say, the decisive aspect about the assessment of technology unfolds again in the opposition between a technological determinism that resounds Carlyle and the promise of a progressive liberation through the technological that echoes Jefferson and Walker. Marx seems to collapse the distinction, however, when he references Fordism and Frederick Winslow Taylor as epitomizing capitalist liberalism: "Taylor's theory of scientific management embodies the quintessence of the technocratic" (ibid., 649). Technology, as Marx contends in "Technology: The Emergence of A Hazardous Concept," is not inherently hazardous; rather it is put to hazardous use by a set of discourses and practices that threaten to mystify it and to assign it a sense of agency and the autochthonous (1997, 965–988).

In "Reflections on the Neo-Romantic Critique of Science" (1978) he explores the scientific as yet another variant of the technological. Once again his interest is meta-critical to the extent that he focuses on the cultural codifications of the scientific between determinism and liberation. The backdrop of his argument is Charles Percy Snow's thesis of the antagonism of two cultures: the natural sciences, endowed with the empirical and the rational, and the humanities, endowed with the imaginative and the ethical. Writing prior to his hire at MIT, Marx constructs a genealogy of institutionalized cultural and literary criticism, arguing that Ivor Armstrong Richards and subsequent New Critics "tended to emulate the posture of the dispassionate, impersonal, scientific observer," which led them to prioritize "the 'how' as against the 'what' or 'why' of literature" (1978, 66). For Marx, such an approach is flawed—if nothing else, because it risks to position the study of literature less "as a capacity of general culture, available to all educated people" but rather as "an arcane skill" of a professionally trained elite (ibid., 66, 67). He contrasts the New Critics's transposition of the scientific and of methodology onto literary studies with the cultural historian Theodore Roszak, who shares with Richards a belief in "the antithetical character of objective and subjective, and therefore of scientific and moral (or aesthetic), modes of thought," but who, in extension to Carlyle, rejects the primacy of parameters of quantification and verification and the rationalization of an entire worldview that permeates society (ibid., 67).[9]

Against the backdrop of both the idea of two irreconcilable cultures and the pervasiveness of the technological in the guise of the scientific, Marx historicizes the twentieth-century debates by looking back to Romanticism and the rapid developments of the nineteenth century. He contends that "the modern romantics (Wordsworth, Coleridge, Emerson)" were still able to envision "an accommodation between the two modes of perception, empirical and transcendental" (ibid., 68). Yet towards "the end of the nineteenth century the partnership of science, engineering, and capitalism" was firmly established in debates about the scientific (ibid., 72). For Marx, an understanding of knowledge as scientific provides "the intellectual fuel upon which an expanding machinelike society runs;" in "the iconography of antiindustrialism," he writes, "machines represent the most conspicuous products, both physical and cultural, of modern science" (ibid., 72). At the same time, as part and parcel of science, machines come to represent both "the analytic mode of thought" and "the social order typified by a perfection of means and a diminished control over ends" (ibid., 72).

9 Marlon Lieber's contribution to this volume complements this line of thought by pointing to a connection between Marx and Herbert Marcuse's distinction between technics (in the sense of techniques of communication, industry, and transportation) and technology as a social process.

The major critique Marx holds against these assessments of the scientific as the technological is their obliviousness to "social structures and processes" (ibid., 69), or, as he puts it in his discussion of Heidegger, their tendency to bracket "the socio-economic or political realm where material forces and social practices interact" (1984, 643). While revisionist scholars would charge Marx's study with lacking a systematic inquiry into these practices, processes, and structures, his mobilization of the social is institutionally significant since it provides the link between the humanities and the social sciences. Moreover, if Marx insists on the socio-economic and the political as the ground where "material forces" (i.e. technological means and inventions) and "social practices" (i.e. the use and utilization of these technologies) collude, this implies an understanding of the role of the practitioner of American Studies as socially relevant precisely because he or she discusses this collusion. If nothing else, this directedness toward the practices and their consequences as the goal of inquiry indicates the legacy of pragmatism, to which we will return at the end of this essay. For now, there remains the task of properly assessing the interplay of the machine and the garden as a cultural dialectic.

The Trope of the Machine in the Garden

The first pages of *The Machine in the Garden* together with the "Afterword," added in the thirty-fifth anniversary edition of 2000, provide a suitable leverage point for assessing Marx's notions of art, culture, literature, nature, and technology. Alluding to and footnoting the work of William Empson, Frank Kermode, Erwin Panofsky and others, he introduces his own distinction between "two kinds of pastoralism—one that is popular and sentimental, the other imaginative and complex" (2000, 5). This distinction is contingent upon another differentiation between high culture and popular culture. While there is clearly a value judgment implied here that has provoked many critics to make an appearance, it would be premature to conclude that Marx's text does not also show an awareness of the constructedness of this differentiation. After all, he consistently puts "high" in quotation marks, indicating an understanding of the artistic and literary canons as constructed entities that precedes the advent of cultural studies, whose proponents charged him with promoting an elitist conception of the arts and literature. He sets apart, though, and thereby clearly privileges "the literary imagination," declaring his interest "to understand the interplay" of this imagination "and what happens outside literature, in the general culture" (2000, 4). By implication, literature is positioned as a specific part of culture, both inseparable from and conditioned by it, a distinct form of imagination produced by a culture that is itself no less imagined. Such navigating distinguishes him from the New Critics on the one side, and it complicates his relation to one of his former teachers, F. O. Matthiessen, on the other, to whom he owes his interest in the literatures of the American Renaissance in particular—an interest that is most visible in the title of the dissertation manuscript which he began in the late 1930s and submitted at Harvard in 1949, and which forms the basis for *The Machine in the Garden*: "Hawthorne and Emerson: Studies in the Impact of the Machine Technology on the American Writer."

The key figure that serves to refine this positioning is once again Leo Marx's teacher Henry Nash Smith. In *Virgin Land: The American West as Symbol and Myth* (1950), Smith famously suggested to study American culture by tracing mythological and symbolical structures across a broad variety of documents and popular texts—symbols that project the myth of a New Edenic Garden upon the American West. Smith established American Studies as a field of inquiry that from its inception has focused on the documents and texts assigned to what we now associate with the notion of popular culture. Marx followed him insofar as the corpus of texts he reads includes essays, letters, magazine articles, and political speeches. But, as he insists in the "Afterword," he thought that Smith "slighted the challenge to that vision—so conspicuous in the iconography I was assembling—posed by the oncoming industrial order" (2000, 373). This establishes his interest in the intrusion of technology into the pastoral idyll and in the resulting problem of reconciling that "powerful metaphor of contradiction" (ibid., 4), which features, among others, in the notion of the technological sublime, a concept initially coined by Perry Miller (1961–1962), another of his teachers, and subsequently elaborated by David Nye (1994). And this also leads back to his attempt to endow the writers of the American Renaissance and of literary modernism with the capacity to develop and demand literary imagination as a privileged site for the complex critique of the challenges of technological development in general and the rise of industrial capitalism in particular.

To reiterate: the reason why Leo Marx has a bias towards the artistic and literary imagination is contingent upon his differentiation of the two kinds of pastoralism. In its popular and sentimental version, he identifies "an expression less of thought than of feeling" (2000, 5). For Marx, literary imagination allows for a critique of these emotions, which have regained prominence among scholars under the recent turn to affect. It exposes the nostalgic yearning for a retreat from industrial and technological progress implicit in these emotions. In general, Marx's distinction between thought and feeling is once more ambiguous, if contingent upon a value judgment about cultural productions. On the one hand, literary imagination provides us with thought structures that expose nostalgia, melodrama and sentiment in popular culture On the other hand, and this goes back to the tension between the humanities and the social sciences, the humanities and their specific imagination provide a necessary antidote to the rational social sciences and introduce the emotive in the first place. In effect, then, neither emotions nor intellect are per se good or bad; their valences have to be reassessed according to the specific context.

In this context, Marx cites two critics of modernity who profoundly inform his understanding of the nostalgic yearning for retreat from society and technology. The first one is José Ortega y Gasset whose *The Revolt of the Masses* (1930) traces "the division of society into the masses and select minorities" (1957, 15) against the backdrop of the Great Depression, bolshevism and the rise of fascism in Europe. Convinced that societal organization is in principle aristocratic, his prototypically modernist analysis oscillates between an anxiety of the multitude, a fear of the alienation of the self and a general cultural pessimism on the one hand, and an affirmative attitude toward the rise of living standards and an increase in intellectual education on the other. Distancing himself from Oswald Spengler's unambiguously pessimistic outlook that technology might hive itself off and "can go on living when interest in the principles of culture are dead," he insists upon a dialectic understanding of culture and technology (ibid., 83). Yet for Ortega y Gasset, the man of the multitude is unwilling to properly understand this link with its implications for the future and instead engages in enthusiastic consumerism. In this context, he characterizes the man of the multitude as the displaced return of a new Naturmensch. And Leo Marx quotes the following passage from *The Revolt of the Masses*:

> The world is a civilized one, its inhabitant is not: he does not see the civilization of the world around him, but he uses it as if it were a natural force. The new man

wants his motor-car, and enjoys it, but he believes that it is the spontaneous fruit of an Edenic tree. In the depths of his soul he is unaware of the artificial, almost incredible, character of civilization, and does not extend his enthusiasm for the instruments to the principles which make them possible. (quoted in 2000, 7–8)

Marx's ensuing commentary is brief and does not explicate the quotation; it once again betrays a form of essayistic scholarship that resists the structure of topic- and thesis-sentenced paragraphs and that is at odds with the systematic probing of assumptions, which characterize the bulk of later and contemporary critical inquiry. He merely characterizes Ortega y Gasset's passage as a "caricature" that exposes the "conception of reality inherent in our sentimental pastoralism"; he reclaims the "industrial Naturmensch" as a pervasive configuration in American culture (ibid., 8). Reading the citation more closely, we can see that "culture," here, is not only inevitably bound up with a notion of civilization that is contingent upon technology, but that the notion of the culture of the multitude crystallized and criticized consists in the mythological, even biblical idealization and naturalization of technology, enabled by the effective erasure of the boundary between nature and technology. The Naturmensch projects his nostalgia for the pastoral upon the naturalization of the technological. Ortega y Gasset's exposure of a nostalgic naturalization of the technological serves Marx to indicate the necessity of analytically separating the two poles, of stressing their mutual constitution and dependency. The threat that civilization poses to culture lies in its ability to disguise its own artificiality. This disguise produces a category of affect—"enthusiasm": the Naturmensch supposedly celebrates and consumes the surficial products but ignores and can no longer identify the "principles which make them possible," as we can read in the quotation above.

The distinction between surface and depth, which Marx introduces with Ortega y Gasset, returns in the second source he mobilizes: Sigmund Freud. He cites at length a passage from *General Introduction to Psychoanalysis* (1920) and follows it with a question he adopts from *Civilization and Its Discontents* (1930). The passage from Freud's *General Introduction* concerns, and these are Marx's words, "the nostalgic feeling we often attach to the unspoiled landscape" (2000, 8). Freud writes about structures of fantasy and takes as an illustrative example "the establishment of 'reservations' and 'nature-parks'," which are supposed to "maintain the old condition of things which has been regretfully sacrificed to necessity everywhere else" (quoted in ibid., 8). Identifying a pervasive economized rationale of inevitable development and progress, Freud goes on to map the phenomenon of these artificially constructed sites of natural retreat onto "the mental realm of phantasy," which, for him, serves analogously as "a reservation reclaimed from the incursions of the reality-principle" (quoted in ibid., 8). For him, structures of fantasy—whether cultural or individual-psychological ones—mark an attempt to preserve the pleasure principle from within the reality principle. Whereas Ortega y Gasset's Naturmensch enthusiastically naturalizes technology, Freud's reservations, while being artificial constructions, recreate a nostalgic longing for the pastoral, which is associated with an irretrievably lost condition of innocence and pure pleasure. Marx goes on to quote from *Civilization and Its Discontents*, where Freud wonders why "so many people have adopted this strange hostility to civilization" (quoted in ibid., 9). While the answer of the psychoanalyst "is that such attitudes are the product of profound long-standing discontent," the cultural critic Marx, who traces tropological structures and their effects, discovers the recreation of "a symbolic landscape" (ibid., 9). If Freud's analysis endows pastoral sentiment with "a pathological coloring, as if symptomatic of a collective neurosis," Marx looks for a complex alternative to the narratives of modernist alienation revealed in both Freud and Ortega y Gasset, an alternative that would both expose and try to reconcile the dialectic of the pastoral and the technological (ibid., 9).[10]

He finds the most powerful expression of this complex reconciliation in the literary imagination, which he reintroduces in passing, implicitly mobilizing once more the notion of *techne*. He writes that the retreat toward the idyllic pastoral implies a "movement away from an 'artificial' world, a world identified with 'art,' using this word in its broadest sense to mean the disciplined habits of mind or arts developed by organized communities" (ibid., 9). As we have seen, Marx's mode of cultural analysis does not readily affirm or reject either the technological or the natural. Artificiality and technology may achieve the status of something to be criticized for their discipline and organization (here the alienation argument certainly informs the rhetoric); but these characteristics of discipline and organization, once they are expressed in the artistic, also provide the necessary means to expose nostalgia and sentimentalism. With its focus on the ambiguities and ambivalences inherent to concepts and phenomena, this mode of inquiry

10 For a discussion that traces the metaphor of nature-parks from Freud's modernism to Jean Baudrillard see Ulla Haselstein (1992).

bespeaks an avid reader of the literature of the American Renaissance, and of Nathaniel Hawthorne's twice-told tales in particular.

Taking the Romanticist canon and extending it into modernism, Marx argues that the artworks of painters such as Charles Sheeler and of writers ranging from Emerson to F. Scott Fitzgerald are characterized by

> the recurrence of a particular episode [...] in which a machine, or some other token of the industrial power, suddenly intrudes upon the serenity of the writer, or a fictional protagonist, situated in a natural, perhaps idyllic setting. The event typically arouses feelings of dislocation, anxiety and foreboding. [...] The sensory attributes of the engine—iron, fire, steam, smoke, noise, speed—evoke the essence of industrial power and wealth, and when set against the attributes of a natural terrain—fecundity, beauty, serenity, and ineffable numinosity—the effect can be spine-tingling. The recurrence of the "interrupted idyll" testifies to the salience of the conflict of meaning and value generated by the onset of industrial capitalism. (2000, 374)

The pairing of a suggestive, lyrical prose with conceptual analysis shows Marx's indebtedness to an idiom that is profoundly at odds with the one used by many of his later critics who are bent on analytically clarifying the questions of methodology for the by then established field of inquiry, demanding greater systematicity (see e.g. Lipsitz 2001; 2005). It is a rather daring claim he makes, indeed, taking "the recurrence" of an episodic instance, the abrupt invasion of some "token" as the starting point and common denominator of an entire cultural and literary history. On aesthetic grounds, his transfer from the recurring episodic instance to a general tropology has rightfully been criticized by scholars, most convincingly perhaps by Winfried Fluck (1973; 1987), who points to the impossibility of distilling a single structural master trope from reading a text such as Mark Twain's *Huckleberry Finn*. Yet there might be another way of thinking about such critical practice. In the wake of new historicism, scholarship also retrieves such anecdotal and episodic instances in order to create different categories and maps of inquiry.[11] Beyond that, the passage quoted above, in referencing "feelings of dislocation, anxiety and foreboding," once more should make us aware that Marx's scope of aesthetic and cultural analysis includes the realm of affect and emotion (in its dialectical relation to men-

11 Stephen Greenblatt positions the anecdote as that which enables the writing of a counterhistory, arguing that "at the anecdote's rim, one encounters a difference in the texture of the narrative, an interruption that lets one sense that there is something—the 'real'—outside of the historical narrative" (Gallagher and Greenblatt 2000, 50).

tal structures and thought), which has gained renewed prominence in the last decade.[12] Marx's interest in the quotation above clearly lies in the representation of the emotional reactions ("feelings" and "the sensory" and "spine-tingling") produced by the cultural tropology he identifies. These emotions, which he partly assigns to the self-reflexive moments represented in scenes of artistic creation and writing, are ones of "dislocation, anxiety and foreboding;" as such, they challenge the nostalgia of sentimental pastoralism. While this seems to identify Marx with that central narrative of modernity which is bound up with notions of alienation and premonition, we have to once again make the caveat that Marx belongs to a category of thinkers who stress "the conflict of meaning and value" brought about by technological change rather than outright condemning technology.

The major interest of *The Machine in the Garden*, though, lies in tracing artistic and literary reconstructions of "the ancient ideal" of the bucolic: Americans, Marx writes, "cling, after their fashion, to the pastoral hope" amidst the profound technological intrusions (ibid., 355). Charles Sheeler's *The American Landscape* (1930) epitomizes and radicalizes this "curious state of mind" (ibid., 355). In Marx's reading, "Sheeler has eliminated all evidence of the frenzied movement and clamor we associate with the industrial scene" (ibid., 356). Consequently, he contends: "This 'American Landscape' is the industrial landscape pastoralized. By superimposing order, peace, and harmony upon our modern chaos, Sheeler represents the anomalous blend of illusion and reality in the American consciousness" (ibid., 356). *The American Landscape*, in this reading, pastoralizes the technological, aestheticizing it as a bucolic landscape.

Framing his revised study with the theoretical discussion of Ortega y Gasset and Freud and his reading of Sheeler, Leo Marx situates himself and the project of American Studies in the larger, transcultural discourses of both modernity and modernism. The pastoral and the technological form a cultural dialectic in this broad sense, and the U.S. becomes the privileged national-cultural site for analyzing the poles that constitute the dialectic. If Ortega y Gasset's Naturmensch exposes the naturalization of technology in the context of capitalist consumption, Freud's nature parks

12 The reception of Raymond Williams (1977) and his understanding of social formations as contingent upon "structures of feeling" has been crucial. For Sianne Ngai, "anxiety" is one among a series of concepts that constitute what she calls "the aesthetics of negative emotions" of "late modernity" (2005, 1).

epitomize the technologization of nature in the context of urban leisure. And Sheeler's painting represents both relations—the naturalization of the technological and the technologizing of the natural—in an attempt at reconciling them that is always already marked as impossible and fraught with contradiction.

Complexity and the Dialectics of Culture

As Winfried Fluck argues in an essay central to this discussion, the project of early American Studies was "to prove that the United States, the new world power and leader of the Western world, possessed a valuable culture of its own" (2009, 70–71). Because of this, Fluck contends, "one had to define culture as deep and condensed expression of a unique, 'specifically American' identity and of exceptional national virtues" (ibid., 71). Fluck insists on the importance of refocusing on the "cross-cultural exchanges" that facilitated and enabled such claims for a distinct national culture in the first place (ibid., 71). And the design of Marx's study with its varied intellectual and literary sources, from Ortega y Gasset to Freud to Shakespeare, may in fact (somewhat against Fluck's own assessment of Marx) provide a surprisingly suitable starting point for such a transnational project. This does not mean to downplay Fluck's argument that early American Studies aimed "to describe American culture as a modern culture with a specific potential for subversion and negation" (ibid., 72); by contrast, following these intellectual trajectories testifies to and substantiates Fluck's extensive claims about American Studies as part of a larger set of modernist theories that identify and analyze conditions of alienation and that explore options for resistance—but also, as we have seen, for strategic affirmation and affirmative commitment.

To bring our brief study of intellectual influences and institutional conditions that constitute Leo Marx's thinking in *The Machine in the Garden* and subsequent essays to a conclusion, we can say that he reclaims the notion of culture as a dialectics, encompassing various other polarities. In a footnote to the article just mentioned, Fluck reminds us of the programmatic significance of a largely forgotten essay. "Literature, Technology, and Covert Culture," which Marx co-authored with Bernard Bowron and Arnold M. Rose in 1957, seeks to bridge the gap between social sciences and the humanities by bringing together behaviorism and cultural tropology. In this essay, the idea of a covert culture is inextricably bound up with repression, carrying the vocabulary of psychoanalysis. In the dialectics of overt and covert culture, which corresponds to the distinction between surface and depth alluded to in the previous section, the covert is associated with "a recurrent pattern of inconsistent or seemingly illogical behavior" (Marx 1988, 127–128). As an inconsistent pattern, covert culture "is not a whole, complete culture that exists beneath the surface of the overt culture" (ibid., 128); rather it stands in a dialectical relationship with and hence gains its meaning through that which is coded as overt, with which it is inconsistent. But Marx also insists that "covert culture is not the same as alienation from culture" (ibid., 128); by contrast, together with overt elements, covert ones constitute the notion of a cultural holism in the first place. Rather than an inquiry into the subcultural and the marginal, Marx and his co-authors conceive of covert culture as "traits [...] exhibited by the large majority of persons in a society if not by all of them" (ibid., 129). To uncover these traits, social scientists can use "projective tests" and studies such as the Kinsey Report on sexuality or the Rorschach inkblot test to disclose perceptive patterns, and cultural historians may create research designs and samples out of popular literature and canonized literary texts.

Thus, in Marx's thinking, the dialectics of overt and covert culture corresponds to a dialectics of the humanities and the social sciences. Thinking culture in contradictory yet potentially reconcilable polarities that constitute one another without themselves containing an essence, Marx in a general sense follows Lionel Trilling, whose essay "Reality in America" (1950) he references in *The Machine in the Garden* and in the introduction to his collection *The Pilot and the Passenger* (1988). Trilling criticizes what he considers "the economic and social determinism of thought" that characterizes Vernon Louis Parrington's three-volume contribution *Currents in American Thought* from the 1920s (1950, 3). Taking issues with the choice of "current" in Parrington's title, he charges him with omitting any substantial idea of "the culture of a nation" (ibid., 9). Culture, in Trilling's brand of liberalism, is not a vague current, "not a flow, nor even a confluence; the form of its existence is struggle, or at least debate—it is nothing if not a dialectic" (ibid., 9). Marx's own project unfolds this dialectical understanding of culture by introducing a set of distinctions, whose poles achieve their meaning only through their counterpart: the pastoral and the technological, or the garden and the machine; the overt and the covert, or surface and depth. Because each individual pole of these dialectical constellations constitutes itself and gains its full meaning only through its respective opposite, Marx's dialectic corresponds to and produces ambiguous valences: the rational and the emotive, or thought and feeling; alienating technological determinisim and progressive technological liberation, or nostalgic longing for nature and regenerative retreat to it; struggle and debate, or contradiction and reconciliation. It is decisive that Marx conceives of the identification of these dialectical constellations and the assessment of their valences as the prime task of scholarship. While the humanities and the social sciences are united by this task, their difference lies in their access to the emotive. And American Studies, by virtue of being in-between the disciplines, is imagined to provide the middle landscape that keeps the polarities in check.

On the one hand, these layers of dynamic, interrelated distinctions produce cultural complexity—a notion that resurfaces time and again in Marx's rhetoric and that is meant to challenge the clear-cut polarities that inform 1950s consensus narratives. On the other hand, if Marx uses the garden and the machine as conceptual tropes and emphasizes that they gain their specific valence only by contextual and historical situatedness, he also joins a pragmatist trajectory of thought. The version of pragmatism at stake here is perhaps best expressed in William James's "What Pragmatism Means" (1907). James, to recall, uses an anecdotal scene of nature (man, squirrel, tree) as an experimental configuration to explain pragmatism's solution to metaphysical problems, where valences and valuations are always relational, contingent upon context and perspective. In Marx's writings, this means a retreat from melodramatic and polarized narratives of culture and a preference for contextual and historical analysis—an awareness, that is, for the contingencies of value judgments.

The Trope of the Machine in the Garden from the Nineteenth to the Twenty-First Centuries

The various contributions approach and explore Marx's trope and argument from very different perspectives and range from the 'beginnings' of industrialization and the market revolution in early nineteenth-century America to twenty-first-century American bureaucracy and digital culture.

In the first essay "The Politics of Imaging the 'Machine in the Garden' in Antebellum Factory Literature," Katja Kanzler turns to two very different nineteenth-century texts, Herman Melville's well-known short story "The Tartarus of Maids" (1855) and a much lesser known sketch from the genre of 'factory-girl' literature, "A Merrimack Reverie" (1840) written by Abby D. Turner. In bringing to bear Marx's trope on two texts that are specifically concerned with the setting of the factory rather than nature or the pastoral, Kanzler explores a dimension that was neglected in Marx's seminal work: the representation and negotiation of socially determined subject positions. Arguing that the juxtaposition of nature and technology to be found in factory literature is both a reflection of and an intervention into "the rapidly shifting social matrix of an industrializing USA," Kanzler proposes to turn to an analysis not just of the images to be found in the texts but, more particularly, to the "imaging" they perform. While Melville's story engages with the advent of industrialism very much along the lines of Marx's machine-in-the-garden trope, Turner's sketch inverts the trope as she foregrounds the (imagined) presence of horticulture in the factory. It is this difference, Kanzler argues, that betrays the narrators' very different social subject positions as well as, ultimately, the different attempts to engage with the discourse of class within the shifting social landscape of antebellum America.

Kirsten Twelbeck's "Eden Refound(ed): Post Civil War Literary Gardening" explores Henry Ward Beecher's *Norwood* (1867) and the various ways in which the "novel contributed to imaging the postwar nation as a future garden." Against his own failing influence and reputation as a public spokesman, the famous minister and abolitionist had written the novel about the model village and community Norwood, in which a number of issues prevalent in postbellum America are negotiated. Decidedly rural, even bucolic, the village comes to represent the ideal New England community, situated in a Marxian middle landscape between corrupting city and primitive wilderness. In the multiracial but strictly segregated society of Norwood, everybody has their 'natural' place and lives together in harmony, a literary fantasy that is supposed to reconcile a nation torn apart by the ideological implications of the Civil War. Moreover, despite the absence of both machines and a discontented labor force, the Norwoodites enjoy a life of leisure and intellectual pleasure, engaging in reading and discussions, which is a far cry from postbellum American reality. Norwood, Twelbeck shows, is a fantasy, constructing a garden without machines, at least without intruding ones: The railroad featured in the novel serves not so much as an interruption of the bucolic harmony than as a way to bring it to the rest of the nation. And just as the railroad helps Americans to experience this pastoral garden, the novel itself, Twelbeck argues, serves as a "recreational trip" for its readers, offering a literary garden, a proverbial paradise of a rural life of pleasure meant to soothe and reconcile the ideological anxieties of a post-war nation.

In "The Dinosaur in the Garden: From Walden Pond to *Jurassic Park*," Susanne Hamscha reads Marx's analysis of Thoreau's *Walden* (1854) alongside Paul Lauter's exploration of Steven Spielberg's blockbuster from 1993. With this, she brings into dialogue not just two texts from different centuries, genres, and media but also two different moments in American Studies as a discipline. On the one hand, her analysis focuses on the ways in which *Walden* and *Jurassic Park* explore the meaning of 'America': critical of contemporary society, both texts seem to propose a return to a pastoralized, more 'natural' state of America as a solution. While *Walden* discusses the intrusion of the railroad into the pastoral(ized) landscape, the intrusion of the dinosaur into *Jurassic Park*'s scenery complicates this dualism as the dinosaur is both, an embodiment of a 'raw' state of nature and also the result of technology harnessed by capitalist greed. On the other hand, with a comparative analysis of both texts Hamscha brings into dialogue Leo Marx and Paul Lauter and with this, two different (historical) perspectives not just on the state of 'America' but on American Studies as a discipline. As Hamscha argues, Marx's reading of Thoreau—to be seen in the context of the Cold War—focuses on the distinct 'Americanness' of *Walden* and hence, also promotes the idea of American Exceptionalism, which served to underline the nation's role as "the leader of the 'free world.'" Paul Lauter, as Hamscha shows, takes up Marx's final statement that "[t]he machine's sudden entrance into the garden presents a problem that ultimately belongs not to art but to politics" and interprets it as a call to 'politicize' American Studies. According to Hamscha, his analysis of Spielberg's blockbuster shows very clearly his interest in the "cultural work" that texts (from *Walden* to *Jurassic Park* and beyond) perform, their reflection of and intervention in what Hamscha calls "cultural production, distribution, and consumption," a position that in the context of 9/11, in which Lauter's essay was published, gains even more significance.

In "Rural Electrification plus Government Power: Leo Marx, the Materialist Dialectic, and *Power and the Land*," Marlon Lieber explores the dialectic between nature and technology, the garden and the machine, as it becomes visible in Joris Ivens's documentary film *Power and the Land* (1940), which arguably both documented and promoted Roosevelt's New Deal program and, more specifically, its attempt to advance the electrification of rural America. Emphasizing Leo Marx's contention that it is not the machine or, rather, technology per se that intrudes into and threatens to destroy the pastoral garden but the specific use to which it is put, Lieber explores this dimension in the film, in which it is precisely the machine—electrification—that is (supposedly) needed to save the garden—rural America. Lieber's apt analysis of the film works out the contrast set up between an old-fashioned rural America, in which tired farming families meet under the light of kerosene lamps, and a modernized version of it, in which electricity and electric devices help to make their life easier. In this 'electrified' utopia, which the film both extensively 'documents' and advertises, electric appliances seem to appear out of nowhere, a move that hides both the costs and efforts involved in this endeavor. Moreover, as Lieber points out, the film emphasizes in a Jeffersonian fashion that despite electrification the land as well as the people and their virtues remain decidedly rural: it is not profit-seeking private investors, 'big business' as it were, but the farmers' cooperative that helps to electrify the 'garden,' thus ensuring a difference to the city, where electrification has led to unchecked, unhealthy industrialization. In this sense, Jefferson's independent farmer and the pastoral garden he works in is made possible by the machine and its benefits in the form of labor-saving appliances. Ironically, Lieber concludes, it is these very machines that make individual family farming impossible in the long run: Being much more efficient and cheaper than human labor, machines threaten to replace costly human labor, introducing what Lewis Mumford calls the "profit motive" into the garden.

Laura Bieger discusses how two iconic pieces of land art engage with a Marxian version of the pastoral from a specifically postmodern perspective in "Putting Machines into Gardens: Walter De Maria's *Lightning Field*, Robert Smithson's *Spiral Jetty* and the Pastoral Imagination." For De Maria's *The Lightning Field* (1977), as Bieger shows, it is not just the art work itself and how it engages with the surrounding land that is important but also the rigidly controlled circumstances under which it can be seen or, rather, experienced. A little cabin, omitted from the official photographs of the *Lightning Field*, makes the visitor's required 24–hour stay in New Mexico's high desert, where it is situated, possible in the first place and, not least because of its reference to frontier settlements, turns the site into a distinctly pastoral middle landscape. Together with the artificial, 'machine-made' field itself, which does not seem to age or change, the landscape turns into "an idealized, mythical space outside of time," offering a romanticized experience of nature's loneliness and isolation which is contingent on the presence of the machine. Installing Robert Smithson's *Spiral Jetty* (1970) into the landscape required a similar 'machine-made' effort, which links the artwork, according to Bieger, to the building of the railroad or the drilling for oil, traces of which are also to be found in the same landscape. In contrast to De Maria's work, which celebrates a remote, 'empty' (middle) landscape, *Spiral Jetty* is surrounded by traces of unsuccessful human endeavors, an "entropic wasteland" that is decidedly non-pastoral. Moreover, this insistence on historicity that becomes visible in the various traces of 'civilization' scattered in the landscape is also a decisive part of the earthwork itself: it is submerged in a lake whose salt sediments continuously affect and change the *Spiral Jetty*, thus becoming a visible sign of an "entropic notion of history" itself. According to Bieger, the artwork, thus, becomes paradoxically not a rejection of the pastoral but its reversal, an artificially created garden, in which, however, "growth leads to stillness and extinction rather than harvest and progress."

Christa Grewe-Volpp shifts the focus to Native American literature in "The Machine in the Indigenous Garden: Leslie Marmon Silko's *Garden in the Dunes*" and probes in which ways Marx's trope of the machine entering an undisturbed pastoral has to be modified vis-à-vis the fact that Marx's pastoral had not, in fact, been undisturbed when the machine entered. Since in many ways "the pastoral was imagined in opposition to the 'Indians,'" the fact that one of the most renowned Native American writers, Leslie Marmon Silko, takes up the trope in her 1999 novel is, of course, highly significant. Grewe-Volpp shows how Silko rewrites Marx's trope to explore the effects that colonialist and capitalist structures and ideologies have had on Native American cultures. In her creation of an indigenous garden, Grewe-Volpp argues, Silko shows that American nature had never been what Euro-American settlers saw as 'uncivilized wilderness' but a space inhabited and cultivated by indigenous peoples. In Silko's account, the garden is a space of transition and hybridity in which the usual Anglo-American hierarchies and boundaries do not hold; moreover, as Grewe-Volpp's analysis shows, Silko's garden in the dunes is not disturbed by technological progress per se but by the forces of colonialism and militarism which set out to wipe out native cultures.

Klara Stephanie Szlezák turns to "Small Town Love on the Big Screen: The Trope of the Machine in the Garden in Hollywood Romantic Comedies" and thus extends the scope of rereading Marx's argument to contemporary film. Against Marx's contention that the popular and sentimental pastoralism which he traces to popular mass culture lacks complexity and depth and thus, provides simply a form of escapism, Szlezák reads the rom-com as intervening in contemporary culture in much more complex ways. Taking *Baby Boom* (1987), *Doc Hollywood* (1991), and *Sweet Home Alabama* (2002) as her examples, she analyzes how the trope of the machine in the garden serves as an essential element of narrative structure, character constellation and character development. More than mere settings, the city as well as the pastoral signify different sets of values which are at odds with each other as the central conflict between the 'antagonistic' lovers-to-be shows time and again. As Szlezák shows in her reading, Marx's trope can be mapped onto the films' central contrast between urban and rural lifestyle and values with 'city life' entering the pastoral, rural garden like Marx's proverbial machine. Thus, she goes on to argue, the trope is utilized to turn a 'mere' antagonism into a much more complex dialectic, i.e., an interaction that emphasizes the intertwining and interdependence of rural and city life and values much more than a mere opposition would do.

Dietmar Meinel continues the exploration of the trope's validity and function in the area of contemporary Hollywood film with turning to the animated film *WALL-E* (2008). In reading Marx and *WALL-E* together, Meinel argues, the trope is in many ways turned upside down. Not only is the natural garden long gone at the beginning of the film, replaced by the littered landscape of consumerist, post-apocalyptic America; it is the machine in the form of WALL-E, the last remaining clean-up robot on earth, that reconstructs the pastoral and, moreover, reintroduces the humans into it by the end of the film. Hence, the machine in the garden, here, does not so much disturb but reconstruct and recreate the pastoral, a twist of the trope that complicates an easy opposition between conceptions of garden and machine, nature and culture and that shows their intricate intertwining instead. Finally, Meinel argues on a metatextual level, it is also the (computer) machine that creates the (digital) garden, again making a neat distinction impossible.

In "The Neoliberal Machine in the Bureaucratic Garden: Pastoral States of Mind in David Foster Wallace's *The Pale King*", James Dorson engages in a Marxian reading of Wallace's postmodern novel, set in the 1980s and posthumously published in 2011. Approaching the novel from the observation that, for Marx, "the pastoral design is not about a specific type of location but rather about a specific frame of mind," Dorson argues that the text counterintuitively presents the discipline and routine work of a bureaucracy such as the IRS as an immersive and contemplative pastoral retreat from the "social and psychological wilderness" created by the frenzy of the information economy and the counterculture's nihilism and mechanical opposition to bureaucracy. Yet, as Dorson shows, this bureaucratic garden is, in the tradition of Marx's complex pastoralism, eventually disturbed by the intrusion of the "quintessential post-industrial machine" in the form of the computer, which, albeit promising a humanization of routine work and more time for self-realization, introduces neoliberal principles of market-orientation and efficiency which not only threaten the employees' jobs but also the pastoralism of their routine work. Ultimately, Dorson contends, the novel does not nostalgically yearn for a pre-computerized bureaucracy but shows how promises of the freedom of a pastoral state of mind are harnessed by economic forces to further their own interests.

Alan Trachtenberg's epilogue, finally, looks back on Marx's study of the machine in the garden as an important founding moment of the emerging field of American Studies. Pointing to the "historical materialism" to be found throughout Marx's acute analysis of major literary voices of nineteenth-century America, Trachtenberg foregrounds the influence of Karl Marx, something which has hardly been considered within the ongoing discussion of Leo Marx's study. It is to be read not just as an analysis of a diverse range of literary texts of the second half of the nineteenth century but as a "literary-historical-cultural writing" in the context of the Cold War era that asks as to consider history not just as "an array of facts" but "a set of relationships and crosscurrents" played out in a landscape, in which "real space [is] melded with imagined and imaginary space." Moreover, he avers, we should not overlook that in Marx's acute analysis of texts, the machine remains the dominant, if not destructive force in the garden. As, in other words, the "capitalist-industrial order in which the machine reigns" remains the dominant (real and discursive?) power, Marx's call that we are in need of new symbols to both imagine and make happen alternative visions of society becomes all the more urgent in the twenty-first century.

***

This collection of essays is based on the workshop "Shifting Paradigms of the Machine in the Garden: The Economies of the Rural, the Pastoral, and the Technological" conducted by the editors during the Sixtieth Annual Conference "Rural America" of the German Association of American Studies in 2013. The publication of the book in a rather short time in order to meet the fiftieth anniversary of Leo Marx's study was a challenge, and it would not have been possible without the help and generous support of a number of people and institutions. First of all, we would like to thank both the participants of the workshop as well as the other scholars who provided their contributions in such a timely fashion, helping us meet deadlines, most of which were set during the busy time of the summer term 2014. We are also grateful to the series' editors as well as Campus Verlag, who helped this collection on its way in a speedy and uncomplicated manner. Moreover, we would like to express our gratitude to the Ernst-Reuter-Gesellschaft of Freie Universität Berlin, without whose generous financial support this book would not have been realized. Finally, we are deeply indebted to Lars-Frederik Bockmann, whose quick and accurate work on the manuscript made the book look the way it does, and to Alexandre S. Dias, who allowed us to use his beautiful photograph of Olafur Eliasson's "Viewing Machine" to grace its cover.

Works Cited

Bowron, Bernard, Leo Marx and Arnold M. Rose (1988 [1957]). Literature, Technology, and Covert Culture. In Leo Marx, *The Pilot and the Passenger: Essays on Literature, Technology, and Culture in the United States*, 127–138. New York: Oxford University Press.

Bryant, John Lark (1975). A Usuable Pastoralism: Leo Marx's Method in the Machine in the Garden. *American Studies*, 16.1, 63–72.

Carlyle, Thomas (1829). Signs of the Times. *The Edinburgh Review*, 49.98, 439–459.

Empson, William (1966 [1935]). *Some Versions of the Pastoral: A Study of the Pastoral Form in Literature.* Harmondsworth: Penguin.

Fluck, Winfried (1987). *Theorien amerikanischer Kultur.* Konstanz: Universitätsverlag Konstanz.

— (1973). Das ästhetische Vorverständnis der "American Studies." *Jahrbuch für Amerikastudien*, 18, 110–129.
— (2009). Theories of American Culture (and the Transnational Turn in American Studies). In *Romance with America? Essays on Culture, Literature, and America Studies*, Ed. Laura Bieger and Johannes Voelz, 69–85. Heidelberg: Universitätsverlag Winter.
Gallagher, Catherine, and Stephen Greenblatt (2000). *Practicing New Historicism.* Chicago: The University of Chicago Press.
Greenblatt, Stephen (1992). Learning to Curse: Aspects of Linguistic Colonialism in the Sixteenth Century. In *Learning to Curse: Essays in Early Modern Culture*, 22–51. London: Routledge.
Haselstein, Ulla (1992). Vom Urwald in die Wüste: Der Naturschutzpark als Metapher der Kulturkritik bei Freud und Baudrillard. *Amerikastudien / American Studies*, 37.3, 393–414.
Heidegger, Martin (1977 [1954]). The Question Concerning Technology. In *The Question Concerning and other Essays*, Trans. William Lovitt, 3–35. New York: Garland.
James, William (1957). What Pragmatism Means. In *Essays in Pragmatism*, Ed. Alburey Castell, 141–158. New York: Hafner.
Kermode, Frank (1952). *English Pastoral Poetry: From the Beginnings of Marvell.* London: George C. Harrap.
— (1966 [1954]). Introduction. In William Shakespeare, *The Tempest*, Ed. Frank Kermode, xi–xciii. London: Methuen.
Kuklick, Bruce (1972). Myth and Symbol in American Studies. *American Quarterly*, 24.4, 435–450.
Lipsitz, George (2001). *American Studies in a Moment of Danger.* Minneapolis: University of Minnesota Press.
— (2005). Our America. *American Literary History*, 17.1, 135–40.
Loomba, Ania (1989). *Gender, Race, Renaissance Drama.* Manchester: Manchester University Press.
— (2002). *Shakespeare, Race, and Colonialism.* Oxford: Oxford UP, 2002.
Marx, Leo (2000 [1964]). *The Machine in the Garden: Technology and the Pastoral Ideal in America.* Oxford: Oxford University Press.
— (1988). *The Pilot and the Passenger: Essays on Literature, Technology, and Culture in the United States.* New York: Oxford University Press.
— (1969). American Studies. A Defense of an Unscientific Method. *New Literary History*, 1.1, 75–90.
— (1978). Reflections on the Neo-Romantic Critique of Science. *Daedalus*, 107.2, 61–74.
— (1984). On Heidegger's Conception of "Technology." *The Massachusetts Review*, 25.4, 638–652.
— (1997). "Technology": The Emergence of a Hazardous Concept. *Social Research*, 64.3, 965–988.

— (1999). Reflections on American Studies, Minnesota, and the 1950s. *American Studies*, 40.2, 39–51.

— (1999/2000). Afterword: The Machine in the Garden. *The Massachusetts Review* 40.4, 483–496.

— (2005). On Recovering the "Ur" Theory of American Studies. *American Literary History*, 17.1, 118–134.

Matthiessen, Francis O. (1941). *American Renaissance: Age and Expression in the Age of Emerson and Whitman*. Oxford: Oxford University Press.

Miller, Perry (1961–1962). The Responsibilities of Mind in a Civilization of Machines. *The American Scholar*, 31.1, 51–69.

Ngai, Sianne (2005). *Ugly Feelings*. Cambridge: Harvard University Press.

Nye, David E. (1994). *American Technological Sublime*. New Baskerville: MIT Press.

Ortega y Gasset, José (1957 [1930]). *The Revolt of the Masses*. Trans. Anonymous. New York: W.W. Norton and Co.

Pease, Donald E. (1990). New Americanists: Revisionist Interventions into the Canon. *boundary 2*, 17.1, 1–37.

Pease, Donald E., and Robyn Wiegman (2002). "Futures." In Pease and Wiegman (eds.). *The Futures of American Studies*, 1–42. Durham: Duke University Press.

Rowe, John Carlos (2002). *The New American Studies*. Minneapolis: University of Minnesota Press.

Smith, Henry Nash (1970 [1950]). *Virgin Land: The American West as Symbol and Myth*. Cambridge: Harvard University Press.

— (1957). Can "American Studies" Develop a Method? *American Quarterly*, 19.2, 197–208.

Sollors, Werner (1986) *Beyond Ethnicity: Consent and Descent in American Culture*. New York: Oxford University Press.

Spengemann, William C. (1984). The Earliest American Novel: Aphra Behn's *Oroonoko*. *Nineteenth-Century Fiction*, 38.4, 384–414.

Steiner, George (1964). Thought in Green Shade. *Reporter*, 31, 35–36.

Trachtenberg, Alan (1984). Myth and Symbol. *The Massachusetts Review*, 25.4, 667–673.

Trilling, Lionel (2008). Reality in America. In *The Liberal Imagination: Essays on Literature and Society*, 3–21. New York: New York Review of Books.

Warren, Kenneth W. (2011). *What Was African American Literature?* Cambridge: Harvard University Press.

Williams, Raymond (1973). *The Country and the City*. Oxford: Oxford University Press.

— (1977). *Marxism and Literature*. Oxford: Oxford University Press.

Wise, Gene (1979). "Paradigm Dramas" in American Studies: A Cultural and Institutional History of the Movement. *American Quarterly*, 31.3, 293–337.

The Politics of Imaging the Machine in the Garden in Antebellum Factory Literature

Katja Kanzler

This essay brings a fundamentally Americanist question to bear on Leo Marx's fundamental piece of Americanist scholarship: What cultural work does the machine-in-the-garden trope perform in literary texts, texts that—as Marx highlighted—emphatically invoke the socio-economic upheavals of industrialization? Rather than asking what the trope means, I am interested in what it does in textual environments that, literally or metaphorically, navigate a protean discourse of class.[1] I want to pursue this question in a reading of two texts that directly engage with industrialization and its machinery, two pieces of literature written in markedly different circumstances—one by an eminently canonical writer of the American Renaissance, Herman Melville, the other by a woman who worked in the factories of Lowell, the period's model industrial town. My reading of these texts aims to draw attention to the ways in which representations of the machine in the garden are perspectivized: While engaging with the juxtaposition of nature and technology, these representations always also work on negotiating social subjectivities—on defining, contrasting, authorizing, critiquing subject positions in the rapidly shifting social matrix of an industrializing USA. In other words, I propose to not only attend to the texts' images of the machine in the garden but also to the imaging that they depict.

The texts with which I will be concerned dramatize this imaging as work that is deeply situated and entangled in other practices of self-fash-

1 The concept of 'cultural work' is, of course, Jane Tompkins's, introduced in her seminal *Sensational Designs* (1985). And while class has only recently become a more prominent item on the American studies agenda, its salience in antebellum US society and culture has been explored by a number of scholars. Amy Schrager Lang poignantly characterizes the developmental stage of class-relations in this period as a "rehearsal" (2003, 17), highlighting how the social and economic transformations of the antebellum decades increasingly challenged the nation's foundational myth of classlessness and powered an emergent language of socio-economic distinction. For a productive complication of the term 'class' in the context of American literary studies, see Lavelle (2012).

ioning, practices which resonate with industrialism's new regimes of social difference. Herman Melville's short-story "The Tartarus of Maids" (1855) constructs a narrator who renders his encounter with industrialism in a rhetoric greatly informed by the machine-in-the-garden trope. By correlating this figurative practice with the notably limited and biased perspective of its narrator—a perspective whose marking laminates class and gender—the text exposes the work of socio-economic self-fashioning enabled by the trope. The sketch "A Merrimack Reverie" (1840), published in the 'factory-girl'[2] magazine *The Lowell Offering*, develops a motif that seems to invert the trope Marx identified—the motif of horticulture in the factory. This motif unfolds much ambiguity in the text which, I will suggest, registers the precarious quality of the magazine's project to establish the 'factory girl' as an affirmative subject position.

"The Tartarus of Maids"

As so many other texts by Herman Melville, "The Tartarus of Maids" employs a tacitly unreliable first-person narrator, whose contemplations and efforts at sense-making seem to furnish the main interest of the short-story. The nameless narrator had been introduced in the text's first part, "The Paradise of Bachelors,"[3] in which he joined a group of London lawyers for a sumptuous feast, articulating the experience as one of 'paradisiacal' bachelorhood. In "Tartarus," he finds himself in New England and, propelled by a combination of "adventure" and business-interests, travels to a paper-mill (674).[4] While the text does not reveal much about the narrator's character, it highlights the vested interest that sends him on this

2 The term 'factory girl' reflects antebellum usage and is meant to denote the social subject-position and its cultural construction, rather than the historical women who worked in New England's factories.

3 David Serlin describes the literary form of the two-part text as "interdependent portraits, hinged together like medieval diptychs, which are locked in a dialectical relationship that maintains narrative control while it questions its own authority" (81). In "The Paradise of Bachelors and the Tartarus of Maids," the narrator grapples with the precise nature of the relationship between the 'paradisiacal' and 'infernal' worlds depicted. While noting their diametrical opposition, he fails to fully comprehend it—a failure that my reading will address in greater detail.

4 Quotations from Melville's short story will be indicated only by page numbers in the following text.

journey and confronts him with sights of industrialism: He is a businessman, "[h]aving embarked on a large scale in the seedman's business" who, facing happily expanding markets, is looking for ways to cut his expenditures by purchasing his paper at the factory rather than "from the wholesale dealers in a neighboring town" (674). The narrator's discursive management of his dual errand—business and adventure—is highly interesting: His account greatly emphasizes the 'adventurous' dimension of the trip—the thrill of the ride, the aesthetics of the landscape, his affective response to the factory-tour he receives—while all but glossing over the business-transactions that take place.[5] His storytelling thus conspicuously works on casting the narrator as a touristic traveler and visitor to the factory, a self-fashioning that masks his deep entanglement in the economy that sustains the kind of industrial hell he encounters.

The short-story, then, presents itself as a travel-report that the narrator delivers of his journey to and through the paper-mill, whose aesthetic contemplations are informed by his efforts to disavow his own complicity in the industrial economy. These aesthetic contemplations notably iterate the machine-in-the-garden trope that Marx described. The narrator figures his experience of industrialism as one of shockingly dissolving boundaries that point to industrial capitalism, as the true 'counterforce'—dissolving boundaries between workers, machines, and commodities; between the factory, the landscape, and the bodies of the female workers; between industrial production, the capitalist marketplace, and human reproduction. While thus identifying the actual force that drives the transformations of the contemporary American socioscape, this figuration also works at naturalizing these transformations, at masking the agency behind them and the narrator's own entanglements in them. There are moments in the text where this discursive project threatens to implode, where the narrator seems to lose control of its ideological operations. These moments in particular expose the situatedness and investments of the narrator's account and imaging of the machine in the garden, opening up a space for the reader's critical reflection.[6]

5 In fact, his narrative does not directly relate the act of his purchasing, while clearly inferring it when the narrator notes that, after his tour of the mill is finished, he is taken to "the first room I had entered, and where the desk for transacting business stood" (678). The actual transaction is conspicuously not narrated.

6 While most critics have conflated the voice of the narrator in this text with that of its (implied) author, Robyn Wiegman's and Graham Thompson's are among the few readings that emphasize the narrator's unreliability. Thompson perceptively embeds this

The narrator's impressions and contemplations of the landscape through which he travels are punctuated by the motifs that Leo Marx discusses in his study.[7] His account renders the point of departure for his journey as a rural idyll, "bright farms and sunny meadows, nodding in early June with odorous grasses," which he leaves behind as he descends into a landscape he depicts in explicit analogy to hell (673): Traversing a gorge that he calls a "Dantean gateway", he eventually reaches a "great, purple, hopper-shaped hollow" that the locals, as he relates, call "the Devil's Dungeon" (673). There, it is a sound that makes the narrator aware of the factory, with the suddenness that Marx identified as characteristic of the motif:

> At first I could not find the paper-mill. The whole hollow gleamed with the white, except, here and there, where a pinnacle of granite showed one wind-swept angle bare. The mountains stood pinned in shrouds—a pass of Alpine corpses. Where stands the mill? Suddenly, a whirling, humming sound broke upon my ear. I looked, and there, like an arrested avalanche, lay the large white-washed factory. (674)

The sound of the machine that suddenly shoots through the landscape does not so much mark—by violating—the distinction between nature and technology in the narrative, it rather accentuates a continuity that the narrative has established throughout its depiction of the narrator's descent to the Devil's Dungeon. The chief contrast in the text is between the rural idyll from which the narrator departs and the landscape he passes as he approaches the paper-mill. This landscape, foreshadowing the 'hellish' conditions at the factory, presents itself as nature already infected and transformed by industrial technology. An old, decayed factory-building that the narrator encounters on his journey suggests that this landscape looks back on a history of industrialization, framing its infernal properties as the marks left by industrial colonization.

short-story in Melville's propensity to use narrators that "offer a version of events that is subtly convincing and yet as interrogative of the narrators themselves and their subject positions as it is of the events narrated" (523).

7 In fact, Marx refers to this short-story, albeit only once, to briefly illustrate the motif of the machine's sudden appearance in the landscape (15). I discuss the passage from the short-story he quotes below.

Significantly, the narrator figures these infernal properties not only through analogies to hell,[8] but layers them with invocations of female anatomy, sexual intercourse, and gestation. Several scholars have commented on this dimension of the text. Robyn Wiegman may count as exemplary in her observations on how the landscape that the narrator describes on his passage to the paper-mill resembles women's reproductive anatomy: He "encounters the landscape of a grotesque female body", Wiegman notes (1989, 742): "Moving from 'bright farms' to 'bleak hills' [...], the narrator penetrates—in the masculine imaginary of heterosexual intercourse—a 'Dantean gateway' [...], a dark hellish entrance into the vagina" (ibid., 742). The "hopper-shaped hollow" at which he finds the factory evokes a uterus, as John Lavelle remarks (2012, 115), and the tellingly named 'Blood River' that drives the factory's machinery conjures up menstruation. The narrator's depiction of the mill's interior, and particularly of its machines, extends on this imaging. He observes a female worker strapped into a phallic machine whose operations evoke sexual intercourse: a "huge frame of ponderous iron, with a vertical thing like a piston periodically rising and falling upon a heavy wooden block" (675). Later, he uses images of gestation to narrate the transformation of raw-material into paper effected by the mill's central machine, depicting it as a process in which "white, wet, wooly-looking stuff, not unlike the albuminous part of an egg, soft-boiled" is converted, in a nine-minute operation, into paper and "delivered" into the hands of a worker who, as the narrator's tour-guide points out, used to be a nurse (676, 677).

I read these figurations as signifying on the narrator's discursive practices, practices that depict the manifestations and effects of industrialization in terms of dissolving boundaries not only between the factory and its surrounding landscape, but also between the natural landscape and the bodies of the women who work at the factory, between the industrial production they are compelled to perform and the reproduction for which their bodies are 'actually' designed.[9] The narrator's practice of thus figuring

8 Young's reading engages with these 'infernal' references in greater detail, highlighting that they not only evoke biblical hell but also the underworld of Greek mythology and Dante's *Inferno* (1991).

9 The short-story's evocation of female sexuality and reproduction has inspired various interpretations. My reading here goes in a similar direction as Konstanze Kutzbach's, who argues that—while the text critiques the exploitation of women and/as workers—it operates on the basis of the essentialist gender-discourse that circulated in mid-nineteenth-century society (2007). The text's figuration of workers as women reflects a

his encounter with industrialism is interesting on two planes. First, in its iteration of the motif of dissolving boundaries, it takes consistent recourse to an image that diffuses questions of agency. Nothing and nobody causes the hell that the narrator experiences. His account, quite literally, naturalizes its occurrence and thus deflects attention not only from the socio-economic power structures that sustain and are sustained by this 'hell' but also from his own involvement in them.

Second, however, the narrator's discursive practices do approximate—in figurative terms—an acute analysis of industrial capitalism and its social impact. It is particularly his image of the worker as indistinguishable, both from the industrial machine and from the commodities she produces, that reflects this latent understanding. The image of the-worker-as-machine emerges when the narrator describes the women's operation of the machines, highlighting the 'mechanistic' repetitiveness of their labor which he places in opposition to the human activity of speech:

> Not a syllable was breathed. Nothing was heard but the low, steady, overruling hum of the iron animals. The human voice was banished from the spot. Machinery—that vaunted slave of humanity—here stood menially served by human beings, who served mutely and cringingly as the slave serves the Sultan. The girls did not so much seem accessory wheels to the general machinery as mere cogs to the wheels. (675)

The image of the workers' assimilation with their industrial environment—whose shocking quality is augmented by the reference to slavery—also resonates in the first impression of the factory's interior that the narrator articulates, a remark that additionally figures in the commodity that the workers produce: "At rows of blank-looking counters sat rows of blank-looking girls, with blank, white folders in their hands, all blankly folding blank paper" (674). The women's 'blankness', their "sheety" appearance, affords a recurrent motif in the narrative, which metaphorically likens the women to the paper they make (676). The image of the-worker-as-commodity as well as that of the-worker-as-machine operate as powerful metaphors of industrial capitalism's degrading effects, metaphors that reflect an almost (Karl) Marxian analysis. The worker-as-commodity tropes what

common practice in the period's literature. This practice, on the one hand, responds to the demographics of New England's early industrial workforce; discursively, on the other hand, it displaces aspects of the newly emerging distinctions of class onto the already naturalized distinctions of gender. For a more detailed discussion of this dynamic also as it applies to Melville's short-story, see Lang (2003) and Dimock (1994).

Karl Marx theorized as the reduction of workers to the labor they can sell, i.e., the reduction of human beings to commodities: "The worker becomes an ever cheaper commodity the more commodities he creates. [...] Labour produces not only commodities; it produces itself and the worker as a commodity" (1978, 71). The worker-as-machine image metaphorically captures Karl Marx's concept of alienation, in ways also discussed by Leo Marx in his reading of Romanticism's anti-industrial rhetoric: Depicting human beings as having become machine-like "means that their behavior is increasingly determined externally, which is to say, by invisible, abstract, social forces unrelated (or alien) to their inward impulses" (1964, 176)—abstract forces that Karl Marx traced to industrial capitalism.

While the narrator's figures of speech thus lucidly reflect on some of capitalism's fundamental dynamics, he never arrives at a full realization of their import. Twice in the narrative, he seems to come close to such a realization but is interrupted by a sudden intervention. The scenes allow for two readings of where these interventions come from: either it is the narrator's (unconscious) self, his body intervening in an interpretive activity that would eventually lead to an understanding of his own complicity in the system that exploits women/workers; or it is a representative of capital who, in each instance, steps between the narrator and actual understanding. In one of these scenes, the narrator stands transfixed before the factory's central machine, and pondering it inspires reflections that—in ways reminiscent of the "technological sublime" Leo Marx theorizes—come close to making a connection between the machinery of industrial capitalism and the workers' pain he has noticed: "Before my eyes—there, passing in slow procession along the wheeling cylinders, I seemed to see, glued to the pallid incipience of the pulp, the yet more pallid faces of all the pallid girls I had eyed that day. Slowly, mournfully, beseechingly, yet unresistingly, they gleamed along, their agony dimly outlined on the imperfect paper" (678). At this point, his appointed tour-guide yanks the narrator away from the object of his critical meditation: "'Halloa! the heat of the room is too much for you [...] Come out, Sir—out—out,' and, with the protecting air of a careful father, the precocious lad hurried me outside" (678). The second scene has the narrator talk to a character named Old Bach—"the principle proprietor" of the mill—in the context of his (narratively unacknowledged) purchasing of the paper (678). Asking Old Bach about the workers and learning that, by company policy, they are all unmarried, the narrator begins to reflect on the celibacy enforced in and by the factory—reflections that seem to revolve around an imaging of the factory as a cloistral-like space in which the women are 'wed' to industrial capitalism. His thoughts are markedly inarticulate: "some pained homage to their pale virginity made me involuntarily bow" and a "strange emotion filled me" (678). In a dynamic that echoes the earlier scene, Old Bach interrupts the narrator's reflections precisely at the point when he seems to be on the brink of understanding: "'Your cheeks look whitish, Sir,' said the man, gazing at me narrowly. 'You must be careful going home. Do they pain you at all now? It's a bad sign if they do'" (678).

These two scenes compellingly dramatize the narrator's approximation of but ultimate failure to reach an understanding of the industrial spectacle he has witnessed, in ways that reflect both on his middle-class subjectivity and on the nature of industrial capitalism. They do so not only by staging these well-timed interruptions, but also by highlighting the narrator's inability to verbally express his realizations. His inarticulateness is even more apparent in the text's storyworld, where the narrator's vision of the workers' faces on the paper that passes through the machine remains unpronounced. To the contrary, what the narrator articulates when he meets Old Bach immediately after the scene is a clichéd compliment on the factory's technology: "Yours is a most wonderful factory. Your great machine is a miracle of inscrutable intricacy" (678). But also on the level of his narrative discourse, the narrator cannot verbalize what he seems to begin to understand. His responses are conspicuously non-verbal—a "bow" and a "strange emotion". The narrator's failing language—troping his inability or unwillingness to fully understand the spectacle of industrial capitalism—culminates in his final words. On his way back up the ravine, he pauses and looks back on the paper-mill, exclaiming "Oh! Paradise of Bachelors! and oh! Tartarus of Maids!" (678). This "oh!", reminiscent of the "ah!" with which Melville has the narrator in "Bartleby the Scrivener" end his account, presents a conclusion to the narrator's experience that draws emphatic attention to its own inarticulateness. The "oh!" marks a latency of understanding as well as an affective response, which the text's ending specifically connects with the question of the interrelationship between the worlds of mirthful affluence and wretched labor that its two parts depict—a question that ultimately involves the narrator's own positionality.

By thus demarcating the limitations of its narrative voice, the text draws attention to the investments that inform the narrator's storytelling. It exposes the situatedness of his engagement with industrialism and frames his responses—which tap into the rhetoric of the machine in the garden—as a specifically middle-class encounter with industrialism. The imagery of shockingly dissolving boundaries that the text adapts from the machine-in-the-garden trope and puts into the mind and mouth of its narrator are framed as his ambivalent confrontation and struggle with the middle class's entanglement in the economy of industrial capitalism.

"A Merrimack Reverie"

In contrast to the fictional workers in Melville's short-story—marked by their silence as well as the blankness they share with the paper they produce—the historical women who worked in New England's early factories were quite vocal and produced a notable amount of writing. Literary magazines "written by factory girls"[10] were a wide-spread phenomenon in the region's industrial towns in the 1840s. The literary projects pursued by these writers were overdetermined by their socio-economic position in antebellum US society. Magazines like *The Lowell Offering* specifically cast the authorial voices they compiled as those of factory girls, and most of their texts—featuring factory-girl characters and/or narrators—explicitly or implicitly address social expectations about industrial workers. Writing, especially in the modes of genteel literature, serves these authors as a strategy to distance themselves from images of proletarian degradation like those articulated by Melville's short-story and circulated even more widely by popular 'exposé' fiction,[11] and to claim inclusion in a category of 'Americans' conceived, in accordance with hegemonic discourse, as classless. The very performance of writing as a factory girl—or, as Sylvia Cook puts it in her study of working women's literature, of "presum[ing] to pursue a life of the mind while performing labors of the body"—engages with the shifting social matrix in the industrializing nation and seeks to carve

10 "Written by Factory Girls" is the subtitle of *The Lowell Offering* (1840–1845), the most renowned venue of factory-girl-literature.

11 There are several sensationalist novels from this period that focus on the suffering of female industrial workers, often by figuring her exploitation through variations of the well-established seduction plot. See, e.g., Osgood Bradbury's *The Mysteries of Lowell* (1844) or J.A.B.'s *Mary Bean: The Factory Girl* (1850).

out an affirmative subject position for the period's (female) industrial worker (2008, 3).

In the following, I single out a particular motif that appears in a notable number of factory-girl texts, a motif that operates within this project of negotiating working women's social position while deeply resonating with the rhetoric that Leo Marx examines in his study: the motif of horticulture in an industrial environment, of potted plants inside the factory. Evoking while inverting the machine-in-the-garden trope, the motif also builds on the juxtaposition of apparently oppositional signifiers of nature and technology. The affective response that this juxtaposition targets—not the irritation of the classic trope but in many ways its opposite—immediately contributes to the overarching discursive project of factory-girl-literature. At the same time, however, the motif also registers the difficult, haphazard quality of this project—to aim at minimizing class distinctions at a time when social and economic developments were rapidly exacerbating them. This precariousness manifests itself in moments of tension or ambiguity in and around the motif of plants in the mill. My discussion will focus on one exemplary text, which I will embed in its textual environment of factory-girl-literature—the sketch "A Merrimack Reverie," written by Abigail, a pseudonym used by Abby D. Turner[12] and published in one of the first issues of *The Lowell Offering*.

Allusions to horticulture in the mills tend to be brief. In "A Merrimack Reverie," they exhaust themselves in a seemingly passing reference at the text's beginning: "During a leisure moment in the mill, I stepped to the window to examine the plants I am cultivating. My eye glanced out upon the noble Merrimack, flowing in silent beauty and grandeur almost beneath my feet. Being in a rather roving mood, I begin to sail up the stream" (29).[13] The remainder of the short sketch tracks the narrator's imaginary journey along the river, past Lake Winnipisseogee and Pawtucket Falls as

12 Pseudonymous publication was customary in *The Lowell Offering*. The former 'factory girl' and contributor to *The Lowell Offering*, Harriet Hanson Robinson, compiled a key to the pseudonyms which Judith Ranta lists in her excellent *Women and Children of the Mills* (1999, 300). Robinson's autobiography is also the only source available for any information about Abby D. Turner. Robinson notes that Turner joined the Lowell workforce together with her sister Elizabeth at the young age that was common for 'factory girls.' She further recalls her as "one of the earliest writers for *The Offering*; her first contribution was written when she was sixteen years of age" (1898, 192).

13 Subsequent quotations from this story will be indicated only by page numbers in brackets.

well as several canals, each of which inspires a set of meditations, before the narrator returns to the diegetic here and now of the factory, where "the machinery in our room was rattling away as merily [sic] as ever" (30). While the reference to potted plants in an individual text like this may seem slight, it receives marked emphasis from its textual environment: In the whole body of factory-girl-literature, horticulture in the mills emerges as a well-carved out theme. Jane Weiss, in her study of the correlation between nature and spirituality in *The Lowell Offering*, surveys a significant number of texts that discuss plants and flowers in the factories (2011). One such text is published in immediate vicinity of Abigail's sketch. The editorial of the issue that contains her text is dedicated to the topic: "We have been greatly pleased with the taste and care displayed in the introduction and culture of plants and flowers, on all the Corporations. [...] it is highly gratifying to see them exalted to companionship in the sitting-room and parlor. [...]. It is especially gratifying to behold them thriving beneath the kindly care of the female operatives in our factory" (Anon., 1840, 32). The intertextual network in which an individual text like "A Merrimack Reverie" is embedded thus greatly accentuates its brief reference to horticulture and allows it to emerge as part of a larger topos.

The motif of plants and flowers in the mills appears to work like an inversion of the machine-in-the-garden trope, staging a 'surprising' intrusion of nature in a setting marked by industrial technology. Also its discursive function seems to invert that of the 'original' trope, aiming to soothe—rather than express and explore—anxieties around industrialization. This soothing, on the one hand, is effected by casting a technologized environment as hospitable to nature. The image of potted plants on the factory's window-sills seems to iterate what Marx described as the idea of the 'middle landscape,' a space in which nature and civilization are reconciled. On the other hand, the reassurance offered by the image of flowers in the mills also speaks to anxieties around class and gender. Horticulture is an activity linked to the performance of class- as well as gender-identity. As the wording of the above quoted editorial illustrates, potted plants and flowers liken the factory shopfloor to the parlor, the period's emblematic space of both femininity and middle-classness. The motif of horticulture immediately targets images of working-class degradation and of industrial labor's dehumanizing effects, allowing texts like Abigail's sketch to distance the factory girl from such images. Claiming the factory as an ersatz-parlor, the text emphatically constructs the subject-position of the factory girl as consonant with the period's ideal of the True Woman.[14]

However, the motif's operation as an inversion of the machine-in-the-garden trope in "A Merrimack Reverie" is irritated by the text's almost complete silence on trappings of industrialism—a feature that the sketch shares with most publications in *The Lowell Offering*. The factory's machinery is addressed only once, regarding its sound—whose depiction as 'happy rattling' appears like a gross defamiliarization[15]—while, as objects, the machines remain unrepresented. The plants are the only objects in the mill's interior that the text discusses, and their care is the only work its narrator is shown to perform. The introduction of an emblem of pastoralism appears to absorb the text's attention, its effort to cast the factory as coextensive with bourgeois domesticity seems to distract from its original project of writing the factory girl and her industrial environment. What the text depicts, after all, is not the reconciliation of the 'middle landscape,' nor is it a subject-position that reconciles the model of bourgeois femininity with wage labor in an industrial setting. It rather eclipses the industrial machine and the labor that is associated with it to facilitate a quasi-escape into nature. This conspicuous absence greatly signifies in the text. It is deeply connected to the motif of horticulture in the mills, whose elaboration in texts like Abigail's sketch entails an erasure of industrial labor and machinery.

"A Merrimack Reverie" apparently advances this erasure by the movement of its plot, which takes the narrator outside of the factory, on an imaginary journey along the river. The narrator's industrial labor is not part of the narrative—the story-time is demarcated as a break of "leisure" that she takes. Yet while this textual move, on one level, further aids the absencing of industrial labor, it also enables the text to address industrialism and its regime of labor after all, to figure it indirectly in a narrative that draws on well-established literary formulas. This indirect figuration begins

14 "True Womanhood" is the term Barbara Welter uses to denote the period's normative ideal of femininity, encompassing the virtues of piety, purity, domesticity, and submissiveness (1966). The motif of horticulture is one strategy by which the magazine's texts stage the factory as a quasi-domestic space, in an effort to approximate the ideology's mandate of domesticity.

15 The Cotton Mills Museum in Lowell periodically sets their exhibited machines in motion to give visitors an impression of the deafening noise they emitted. The debilitating noise is one of the most frequent items of complaint in texts by Lowell's workers in this period.

with the text's adaptation of the travel-plot. The narrator's journey is not real in the text's storyworld but, as additionally highlighted by the text's title, only takes place in her imagination. The imaginary quality of her journey, which markedly distinguishes her narrative from the numerous travel-narratives that punctuate antebellum literature, highlights her immobility, a confinement that resonates with the industrial regime of labor, while expressing a deep desire for mobility, for escape. This desire particularly registers in the language of regret through which the narrator relates even her imaginary return to the factory: "Being obliged to leave this sweet and quiet spot, and return to the city of spindles, I began more leisurely to retrace my course" (29). As a fantasy, the travel-narrative brings the factory back into the text and expresses a discontent with industrial labor that stands in marked tension to the text's surface.

A second dynamic by which the text indirectly figures industrialism concerns its development of the potted plant as a transcendental gateway: Examining her plants sets in motion the narrator's imaginary journey, a series of associations and reflections that move from the plants to the river on the other side of the window, and along the river's natural as well as 'man-made' course. At the beginning, the trajectory of these reflections seems to aim at a didactic endorsement of industrialization: Passing various natural phenomena along the banks of the river, culminating in a lake whose 'wilderness' is accentuated by the narrator's comment on "the beautiful name given it by the Indians" (29), the narrative arrives at a waterfall harnessed in the service of industry: "Here I again paused, not as before to admire the beauties of nature alone, but to observe also the wonderful works of art in dams, and canals, to raise, receive and carry the water through almost every part of the city, and then causing it to put in motion such a vast quantity of machinery, by which thousands of persons are profitably employed" (29). The passage raises the expectation that this would be the point of the narrator's return, that her narrative would follow up on the apparent resonance between the harnessed waterfall and the machines she operates to bring the factory girl back to the factory. The image of horticulture that stood at the beginning of this associative chain—emphatically an instance of cultivated nature—seems to anticipate such a gist of the plot. However, this is not what the text delivers. Its narrator both delays her metaphorical return to the factory and directly challenges the industrialist euphoria the text had seemed to invoke: "while our admiration is deeply excited by the contemplation of all these noble works of art, and the readiness with which the powers of nature are made subservient to the use and benefit of the community, we cannot but be forcibly struck with their inferiority, when compared with the sublime scenes of nature" (29). The text immediately proceeds to illustrate this hierarchy: While technologized nature provides for 'profitable employment,' the sublime wilderness enables the text's reflections on god and human nature. In thus juxtaposing wilderness and technologized nature, the sketch not only values the 'work' of philosophical and spiritual contemplation over that of waged labor, it also dramatizes their difference in literary productivity. The meditations inspired by 'uncivilized' nature provide for the bulk of Abigail's sketch, a textual presence that throws into relief the absence of 'profitable employment' as a subject of her writing. The text here confronts the logic by which hegemonic discourse in the period reconciles the machine and the garden, as Marx argues, in a rhetoric of progress, emphasizing rather their distinction from the subject-position of an industrial worker.[16] Again, the tacit protest articulated here rubs against the note of cheerfulness with which the sketch ends, the strange image of the machines' 'cheerful rattling.'

Such moments of tension or ambiguity coalesce around the motif of horticulture in the mills—an effort to reconcile the machine with the garden that ends up erasing the machine from the surface of the text; a retreat from the factory that ends up re-inscribing the experience of industrial labor into the narrative; and an endorsement of "the ideology of American industrialism" that ends in its challenge (Marx 1964, 158). I read these moments as symptoms of the discursive problems that the project of genteel factory-girl literature confronts. The hegemonic discourses that these texts try to adapt—in order to fashion female workers into classlessly 'American' True Women and to cast the factory as coextensive with the True Woman's domestic habitat—resist such adaptation. As in Herman Melville's very different text, images of the confrontation between nature and industrial technology emerge as deeply situated productions in the sketch, serving as a platform for the conflicted negotiation of social subjectivities. These two exemplary readings thus alert us to the perspectivized

16 Such glorifications of 'wild' nature vis-à-vis cultured or technologized nature is a recurrent theme in factory-girl literature. As Chad Montrie argues, this theme constitutes a key discourse by which workers express their "resistance to mill labor" (2008, 14), articulating their discontent with a regime of labor that entailed "routine confinement among clattering machinery and noxious lamp smoke" (ibid., 22).

work of imaging that underwrites literary representations of the machine in the garden and to the complex politics they unfold.

Works Cited

Abigail [Abby D. Turner] (1840). A Merrimack Reverie. *The Lowell Offering*, 1.2, 29–30.

Anonymous (1840). Editorial: Plants and Flowers in the Mills. *The Lowell Offering*, 1.2, 32.

Bradbury, Osgood (1844). *The Mysteries of Lowell.* Boston: E.P. Williams.

Cook, Sylvia Jenkins (2008). *Working Women, Literary Ladies: The Industrial Revolution and Female Aspiration.* Oxford: Oxford University Press.

Dimock, Wai Chee (1994). Class, Gender, and a History of Metonymy. In Wai Chee Dimock and Michael T. Gilmore (eds.). *Rethinking Class: Literary Studies and Social Formations*, 57–104. New York: Columbia University Press.

J.A.B. (1850). *Mary Bean: The Factory Girl. A Domestic Story. Illustrative of the Trials and Temptations of Factory Life. Founded on Recent Events.* Boston: Hotchkiss.

Kutzbach, Konstanze (2007). The Two-..., One-..., None-Sex Model: The Flesh(-) Made Machine in Herman Melville's "The Paradise of Bachelors and the Tartarus of Maids" and J.G. Ballard's *Crash*. In Konstanze Kutzbach and Monika Mueller (eds.). *The Abject of Desire: The Aestheticization of the Unaesthetic in Contemporary Literature and Culture*, 181–196. Amsterdam: Rodopi.

Lang, Amy Schrager (2003). *The Syntax of Class: Writing Inequality in Nineteenth-Century America.* Princeton: Princeton UP.

Lavelle, John F. (2012). *Blue Collar, Theoretically: A Post-Marxist Approach to Working Class Literature.* North Carolina: McFarland.

Marx, Karl (1978). Economic and Philosophical Manuscripts of 1844. In Robert C. Tucker (ed.). *The Marx-Engels Reader*, 2nd ed., 66–125. New York: Norton.

Marx, Leo (1964). *The Machine in the Garden: Technology and the Pastoral Ideal in America.* New York: Oxford University Press.

Melville, Herman (1855). The Paradise of Bachelors and the Tartarus of Maids. *Harper's New Monthly Magazine*, 10, 670–678.

Montrie, Chad (2008). *Making a Living: Work and Environment in the United States.* Chapel Hill: University of North Carolina Press.

Ranta, Judith (1999). *Women and Children of the Mills: Annotated Guide to Nineteenth-Century American Textile Factory Literature.* Westport: Greenwood.

Robinson, Harriet Jane Hansen (1898). *Loom and Spindle: Or Life among the Early Mill Girls; With a Sketch of the Lowell Offering and Some of Its Contributors.* New York: Thomas Y. Crowell.

Serlin, David Harley (1995). The Dialogue of Gender in Herman Melville's "The Paradise of Bachelors and the Tartarus of Maids." *Modern Language Studies*, 25.2, 80–87.

Thompson, Graham (2012). The "Plain Facts" of Fine Paper in "The Paradise of Bachelors and the Tartarus of Maids." *American Literature*, 84.3, 505–532.

Tompkins, Jane (1985). *Sensational Designs: The Cultural Work of American Fiction 1790–1860*. New York: Oxford University Press.

Weiss, Jane (2011). "In the Mills, We Are Not So Far from God and Nature": Industrialization and Spirituality in Nineteenth-Century New England. *Journal for the Study of Religion, Nature, and Culture*, 5.1, 82–100.

Welter, Barbara (1966). The Cult of True Womanhood: 1820–1860. *American Quarterly*, 18.1, 151–174.

Wiegman, Robin (1989). Melville's Geography of Gender. *American Literary History*, 1.4, 735–753.

Young, Philip (1991). The Machine in Tartarus: Melville's Inferno. *American Literature*, 63.2, 208–224.

Eden Refound(ed): Post Civil War Literary Gardening

Kirsten Twelbeck

The Civil War transformed American society profoundly on both the political and legal level as it secured the union, brought an end to slavery, and established basic political rights for African Americans. With these changes new problems emerged: early Reconstruction was marked not only by tumultuous scenes and clashes in Congress but also by a culture of denial. As David Blight has famously shown, sectional reconciliation was purchased through the marginalization of black veteran soldiers and by downplaying that the moral crusade against slavery was an important aspect of the War between the States (see Blight 2001). At the same time, the war set in motion a "mental adaptation process" (Butler 2006, 173) that was linked to the profound structural changes that governed American lives:

> [T]he experience of the Civil War shaped the self-reliant nature of the American into a character more amenable to serving in large, hierarchic institutions. Total war spawned powerful, national systems of government, transportation, and industry that operated under centralized control, replacing former localized and individual power bases. When the war began, states issued their own paper currency; citizens and militias owned their allegiance to states. By war's end both North and South had a national paper currency, a national definition of citizenship, and national conscription. The national military organizations served as models for large corporate structures of the future. [...]. The South was destroyed, and the Northern ideal of the independent craftsman, businessman, or farmer was eroded in the process of trying to assert it. Hierarchy, subordination, and obedience to political and economic institutions became the new American way. (Huddleston 2003, 5)

It was during this phase of national transformation and individual adaptation that Henry Ward Beecher, the famous abolitionist, Congregationalist minister, and reformer, became—as Clifford Clark mentions in the subtitle to his biography of Beecher—the "spokesman for a middle-class America" (Clark 1978). His rise to the status of the "most famous man in America" (which is the title Debby Applegate chose for her year 2007, Pulitzer Prize winning book about Beecher) began in 1865, when President Lincoln asked the then fifty-three-year-old preacher to deliver a speech at the raising of the Union flag at Fort Sumter. Having seen the devastation along the route of Sherman's regiment, Beecher was the first to find words of consolation to the Southern population, thereby ushering in a new era of national reconciliation. When Lincoln was shot shortly before the minister's return to New York, Beecher's speech gained political weight. But by 1867 his influence was slowly deteriorating (Stout 2006, 165–178). His contradictory, wavering views were typical of the lack of orientation that ruled among his contemporaries, but he was expected to know better. Particularly confusing were his views on race. In the fall of 1865, Beecher's wife Eunice reported that Henry had claimed before a Charleston audience that "God was black," that Christ was "a mulatto," that the "Devil was white," and that he, Beecher, regretted he had not been born black himself.[1] In February 1866, however, the minister supported President Johnson's vetoing of the Freedmen's Bureau bill that advocated Black suffrage and citizenship. At the same time, he promoted moderate reform in the Southern states—an act that required him to differentiate his position from that of the president, whose ideas of states' rights and minimal federal involvement in the South he had publicly supported. In reaction to this criticism, he tried to establish himself as a moderate Republican, yet this move only spurred a new wave of attacks (see ibid., 176). By 1867 the famous speaker from Fort Sumter had become a scapegoat for the press and was "forced to concede that he could no longer influence the reconstruction policies of the federal government" (ibid., 178). This, then, is the situation out of which arose *Norwood,* a harmonious fantasy that helped stabilize his role as a moral guiding figure. For all its woodenness and thick didacticism, the book was the instant bestseller that Beecher's publisher, Robert Bonner, intended it to be when he asked the famous minister to write it.

Instead of lamenting the novel's obvious and often criticized aesthetic and logical flaws, this article takes the many positive reviews of *Norwood* as its starting point in order to show how the novel contributed to imaging the postwar nation as a future garden. William Dean Howells, for instance, praised the book's optimism in an 1867 review for the *Atlantic Monthly.* It was, he argued, "perhaps the most hopeful element in New York" and

1 Eunice's public statements are quoted in a letter by Calvin Stowe to Beecher, dated October 28, 1865. See Clark 1978, 172.

"the beginning of a social rather than a religious regeneration." In Howells's opinion the novel was "American and good," possessing "sound sense and wholesome impulses" (Howells 1867, 637). Another unnamed reviewer classified *Norwood* as a family or community *event* when suggested that "the true way to enjoy 'Norwood' [was] to read it aloud in the home-circle" (Anonymous 1867, 5). The novel, in other words, was a welcome object for stabilizing mid-Victorian ideals of moral health, family, and community. This article reads *Norwood* as a 'recreational trip' designed to counter the ills of a still deeply divided, anxiety-ridden society. Set almost entirely in a picturesque New England village, the eponymous Norwood, the book celebrates a rural lifestyle far removed from heated government debates and from the scenes of wartime destruction found in many parts of the former Confederacy. By focusing on the representation of New England as national garden, this essay analyzes how Beecher brings together his critique of urbanization with a consoling vision of national progress, inter-regional connectivity, and a highly structured social order even as he actively acknowledges the mourning and suffering that overshadowed American postwar lives (see Faust 2008). Taking a close look at the novel's uses of geographical imagery, this article shows how *Norwood* sought to shape not only postwar American minds but also the nation's immediate landscape. The essay ends with a brief look at a female writer's critical remapping of this profoundly nostalgic landscape. Viewed against the backdrop of Beecher's fantasy of America as harmonious garden, Mary E. Bradley Lane's *Mizora* (1889) emerges as a radical, feminist alternative set in a surprisingly similar yet ideologically even more troubling utopian landscape.

One of the *Norwood*'s central ideological strategies is its invocation of the republican era and American romanticism. Signaling a 'Northern' and therefore 'natural' agenda ("Nor" + "wood"), the book's title reveals its transcendentalist New England heritage. Like many young Americans of the postwar generation, the novel's hero, Barton Cathcart, is tortured by doubt, insecurity, and anxiety vis-à-vis an unclear future. Barton, however, goes through this crisis *before* the war, and he knows where he can find spiritual wholeness and self-reliance; on his strolls in the forest "a bird's call at sundown" gives him "an intuitive apprehension of God's love for him and by this means 'a conception of *infinite* love'" (Smith 1974, 62–63). By highlighting this central moment of conversion, Beecher reminds young readers of America's antebellum cultural resources, including models of meaningful American education: his characters engage in lengthy discussions that bear the stamp of Emerson's "The American Scholar" (1837) and other transcendentalist writings. Apart from that, the title of this novel, its didactic, domestic credo, and its tone all echo Sarah Josepha Hale's *Northwood* (1827, rep. 1852), a highly popular antislavery novel that advocated sending American blacks to Africa to secure a new beginning of an all-white nation.[2] As this article will discuss in more detail later, *Norwood* is not too far removed from this prewar solution.

And yet there are crucial differences between *Northwood* and *Norwood*. Unlike *Northwood*, Beecher's novel is set almost entirely in the North, in New England and promotes a multiracial albeit strictly segregated society. Spanning the time before, during, and after the Civil War, the book represents a village with the name of Norwood and its surrounding gardens and farmlands, as a representative, idealized nation. In line with Leo Marx's definition of a "middle landscape," this "picturesque rather than grand" (2000, 1) environment is "located in a middle ground somewhere 'between,' yet in a transcendental relation to, the opposing forces of civilization and nature" (ibid., 23). "[S]pared the deprivations and anxieties associated with both the city and the wilderness" the novel's protagonists enjoy a way of life marked by family values, a strong sense of community, and an unchallenged acceptance of social hierarchies and divisions. After supporting the Union effort in the new roles of nurse and soldier, white women and blacks happily embrace their presumably natural positions as psychological helpmates and menial servants while Northern and Southern men enjoy lively political and philosophical debates among equals. In keeping with this "cosmic success story" (Smith 1974, 58), the novel follows a strict reconciling agenda: shot through with cross-sectional romance that climaxes with a dashing Virginian's wartime death in a Northern nurse's arms, *Norwood* insists on the superiority of Northern values while paying tribute to the suffering South.

Significantly, it is the New England landscape itself—displaying a "greater variety of scenic beauty than any other that I remember in America" (2)—that enables this fantasy of union, and gives it a quasi-natural grounding. In a highly symbolic move, *Norwood* places a prototypical Southern mansion at the center of a New England middle landscape, thereby contributing to what Susan-Mary Grant has called "the inclusive

2 The novel condemns slavery as un-Christian and damaging to both slave and master. The subtitle of the 1852 edition of *Northwood* was *Life North and South*.

nationalism of the North" (2000, 154).[3] Anticipating the prototypical home of late-nineteenth-century plantation romances, this mansion is inhabited by the benevolent Doctor Wentworth, his angelic daughter Rose (an adult, postwar version of Harriet Beecher-Stowe's Little Eva), and a faithful 'negro.' It is these three, and their role in the novel's miniature nation, that determines the ideological development of *Norwood* and its vision of the future nation.

In order to understand the roles of these three characters we must distinguish Beecher's idealized geography from the hierarchical structure of American plantation culture. Norwood's grand Southern-style mansion is not the jewel on top of an otherwise inferior environment but a sort of entry gate to an extremely balanced, Edenic landscape. Created with the help of neither slaves nor machines, the world of Norwood is unabashedly artificial, a vision rather than a realistic portrait of New England. In line with this concept as literary fantasy, the reader is led (by the narrator) toward the aforementioned mansion "through an avenue of elms." It is in the course of this imaginary, spatial experience that "our homely story" discloses itself (5). The New England that unfolds behind the mansion is a harmonious, simple, and nostalgic ensemble, complete with quaint settlements, well-maintained agricultural areas, and lush forests. "Since the introduction of railways," the narrative informs us, this pastoral fantasy land has been the point of destination for "thousands of curious travelers" who "have thronged New England" (1) to admire the peace and order of this pastoral gem. Suffice it to say that—in keeping with Emerson's vision of the "unifying powers of technology" and his belief in the railroad as "carrying America straight to the pastoral ideal" (Kasson 1977, 122)—this technological interconnection suggests that the 'spirit of *Norwood*' can be both experienced on site and exported—via word of mouth but also in book form—to other parts of the country, if not beyond. Contrary to Leo Marx's famous view of the machine as powerful interruption of a peaceful pastoral, the railroad becomes an integral part of a national geography that relies on the rural as "brood combs" (5) of an idealized state of innocence and friction-less harmony. Confirming what Leo Marx has said about America's antebellum obsession with the railroad, what is announced in *Norwood* is a "new sort of fate" that remains, in keeping with the novel's

3 Grant refers to early reconciliation novels like William DeForest's *Miss Ravenel's Conversion from Secession to Loyalty* that seek to merge what is best in both cultures. Contrary to DeForest's novel, *Norwood* incorporates certain Southern 'traditions' into northern ones.

soothing message, "confined by its iron rails to a predetermined path" (2000, 191).

Generally speaking, *Norwood* turns what Marx has described as the American myth—regeneration in a garden that can be accessed by a train ride (see ibid., 228, 238)—into an experience that can be had simply by reading about it. The novel, in other words, presents *itself* as an alternative to urban, postbellum America by staging itself as a recreational realm that is equivalent to an actual weekend trip to the countryside. From its opening pages *Norwood* takes up the cultural fad of postwar rural tourism, casting the reader as visitor to an imaginary "landscape of the psyche" in Marx's sense (ibid., 28) where communal bonds are intact, thriving, and perfectly suited to evoke a "brimming future" (Wood 1991, 37). The textual village, in other words, suggests a facile shift "from an age of reform to an age of complacency, from an age of egalitarian simplicity to an age of conspicuous consumption and the leisure class," but it also provides a refuge from the political and social tensions of early Reconstruction American (McLoughlin 1970, xi–xii).

Interestingly, the glossing over of that profound rupture that the Civil War brought about accompanies an ideology that promotes acceptance of more rigid structural control and hierarchical organization within the newly centralized, postwar nation that has been described earlier. What this 550-page-long "homely story" (5) suggests is that by embracing these new structures, Americans will reconnect with their past ideals, perfect them, and make an even happier future.

For all its obvious nostalgia, then, *Norwood* is a decisively post-Civil War novel that celebrates the presumably agrarian "brood-combs" (5) of a glorious "Second Founding," a term used by Radical Republicans to emphasize the republican roots of America's post-1865 democratic renewal, and the legacy of the 'city upon a hill'. As the historian David Quigley has pointed out,

> [b]ack in 1787, America's first founding had produced a constitution profoundly skeptical of democracy. James Madison and his coauthors in Philadelphia left undecided fundamental questions of slavery and freedom. All that would change in the 1860s and 1870s. Decided at this second founding were the rules of the democratic game. Though lasting only a few short years, Reconstruction involved countless Americans fighting over who would be able to play in that game, and on whose terms. A century and a quarter later, the democracy that emerged at Reconstruction's end remains our inheritance. (2004, ix)

Norwood participates in this rhetorical battle by projecting an ideal village that is all but isolated from modernity: located on a middle ground "between the rugged simplicity of mountain towns and the easier life in the cities" (4) it represents a veritable postwar pleasure fantasy based upon the antebellum ideal of an "ideological fusion of technology and republicanism" that John Kasson attests to the planners of prewar manufacturing towns (1977, 85). Driven by "their fathers' sense of republican mission and distrust of aristocratic Europe" with its "oppressed, ignorant, and debauched working class" (ibid., 61), American prewar manufacturers sought to supplement the nation's "agrarian commitment" by building factories in rural areas where—in line with what Kasson calls "the new republican industrial order"—the "purifying influences of nature" and "the beneficial discipline of the factory" would go hand in hand (ibid.).

In *Norwood* of course, factories are called "manufacturing villages" (1), and form a constitutive part of an essentially *rural* landscape that merges republican values and a democratic compromise after slavery: in line with the more general ideal of a middle landscape this very conservative republican environment promises *more* participation *without* full citizenship rights. The Norwoodites merge the notion of individual self-realization with the eighteenth-century ideals of social service, industry, and frugality, leaving aside, significantly, the notion of restraint that was so central to realizing a "harmonious society, whose members were bound together by mutual responsibility" (Kasson 1977, 4).[4] Mirrored on the very material level of the novel's middle landscape, this imaginary garden resembles a harmonious mixture of "manufacturing villages" and neatly kept fields and forests—a Jeffersonian ideal that derives its moral legitimacy from its presumably direct contact with nature (see ibid., 19). This vision ignores the technological developments that set in around this time with enormous speed, replacing more traditional methods of farming with improved technology and machinery while traditional manufacturing made room for processes of standardization and mass production (see Hounshell 1985, 167). The novel's denial helps connect the Second Founding with the first one, lending legitimacy, moral grounding, and a sense of historical continuity to the essentially conservative prospect that is sketched out in *Norwood.* Interestingly, the frugal Norwoodites never seem to actively cultivate their agricultural surroundings by either manual labor or with the help of

4 See also Kasson 1997, 18–19, and Marx 1964, 15 and 204.

those improved machines that David Hounshell has described in his chapter on the McCormick reaper and American manufacturing technology. Rather, thriving on the proverbial inventive skill of its Yankee heritage, the community seems to spend most of its time in the village library and its adjacent garden engaging in lively debates about the norms and values of their happy little republic.

Beecher, it seems, wanted to exclusively concentrate on healing the wounds of the Civil War instead of overburdening readers with worrisome speculations about the costs of modernization, acceleration, and social transformation. Inhabiting a radical fantasy of a leisurely, enlightened community, the Norwoodites harvest the fruit of their traditional thrift in the absence of shrieking machines and discontented laborers that were a very real aspect of the age. Only by ignoring the decreasing need for skilled workers and the increasing pressure caused by a growing workforce that preoccupied thousands of Americans at the time can the novel fully celebrate a way of life that it identifies with New England. Class divisions among the Norwoodites do exist, but they are only represented to emphasize the human diversity that makes the village a site of individual self-realization: everybody here makes the most of his or her individual talents and capabilities, and happily contributes those to the harmonious, social organism of Norwood. The novel fashions the region as the Second Founding's model 'city upon a hill' since "no where else in the nation are men so differentiated," so "original, contrasted, and individualized in taste, manners, and opinions" (1). Miraculously freed from the drudgery of work, this New England "*species*" (1, emphasis in original) engage in quasi-Emersonian exchanges among "thinking" men who rely on books as mere inspiration, not instruction, to cultivate an "active soul" (Emerson 1965, 67). Individual, active reading, not manual labor or machines, lies at the heart of the new postwar republic.

It is in this context of a garden without machines that *Norwood* represents a new type the leisurely farmer. Contrary to the traditional concept of the peasant as a worker of the land who is excluded from the increasingly leisurely lifestyle of post-Civil War urban dwellers, Abiah Cathcart enjoys the 'good life' of a 'cultured' person because he knows how to effectively manage time:

> He loved to take his book (he was ever and always a reader) at noon, after his frugal meal was done, and sit by his team, while the horses ground their oats, or

cracked and crunched their Indian corn. Do you wonder, reader, at such pleasure? Then you know little of some scenes of life. (8)

The image of the reading farmer is an old favorite of Beecher's; he sketched it out first in his *Plain and Pleasant Talk about Fruits, Flowers and Farming* (1859), which contains a chapter on "Educated Farmers" as well as the "Portrait of an Anti-Book Farmer." Yet while those texts were written to promote scientific methods in agriculture, and while in the 1850s he had also warned against the threat posed by "evil" books, plays, and other amusements (Beecher 1844, 182), in *Norwood* Beecher depicted a self-aware and disciplined yet leisurely way of life that did not preclude the reading of fiction. (*What* the farmer reads is significantly left open: contrary to Beecher's earlier concern with 'dangerous' literature the farmer may in fact be reading a novel).

More importantly still is the way the protagonist is staged: in keeping with the novel as recreational experience, this scene—which is one of the book's most memorable ones—turns the farmer into a constitutive element of the landscape, thereby producing a harmonious tableau vivant of a pastoral scene.[5] Far removed from the daily bustle of American postwar lives, it advertises the healthiness and beauty of reading as a tranquil alternative to the urban entertainment industry that Beecher himself by that time had learned to enjoy (Applegate 2007, 354) but that resists flexible integration into an individual's working day.

The image of the elderly farmer who contently harvests the fruit of his own frugality may be a figure of the past, but its purpose is clearly future-oriented: with his inner peace and contentedness, Abiah could not be further removed from the war veterans that in 1867 made up most of this older generation, among them many men who never found inner peace and salvation in postwar society (see Faust 2008). It is all the more important, then, that he is staged as a blatantly artificial figure, a sentimental, antebellum ideal, exhibited for the sole purpose of being meditated upon.

Few readers today understand how strongly scenes that call for contemplation rather than attentive reading respond to the psychological needs of a society suffering from memories of violence and destruction. Even William G. McLoughlin, the one scholar who has written extensively about *Norwoood*, considers it mainly a "key work" for anyone interested in

5 Augustin Daly's stage version of the novel relies heavily on the tableau as a memorable scene at the end of each of the play's four acts.

the mid-Victorian (1840–1870) American mind (1970, xi–xii), a negotiation of that "great shift from Calvinism to Liberal Protestantism, from rural to suburban living, from transcendentalism to Social Darwinism, from belief in the omnipresence of the average man to the hero worship of the Horatio Alger captain of industry" (ibid., 6).

In a more immediate context, however, *Norwood* also responded to the traumatic imagery of the war that lingered on for decades after it ended. Only one year before *Norwood*'s publication, Alexander Gardner had published his *Photographic Sketch Book of the Civil War* (1866), a two-volume opus that combined fifty photographs of camps, battlefields, and postwar rubble with text (written by himself) that either described or commented on the images. While the expensive *Sketch Book* itself was a commercial failure, its individual pictures came into wide circulation, defining, as Anthony W. Lee reminds us, "how viewers, then and in subsequent generations, came to know the Civil War" (2007, 1). Set against the iconographic photographs of blood-soaked battlefields, dead corpses with grotesquely dislocated limbs, and heaped up human skulls and bones excavated by a postwar "burial party,"[6] *Norwood*'s tranquil dreamscapes symbolically replace the killing grounds of Antietam and Gettysburg.[7]

Yet *Norwood* advocates a pastoral retreat not only from the immediate past but also from the present. Even in the postwar North, Americans could not escape the daily sight of amputated and mutilated veterans whose injuries were also meticulously documented in medical photography and other visual forms. As Silas Weir Mitchell, who worked on nerve injuries during the war, asserted in 1864, Civil War survivors would carry the war with them for the rest of their lives, suffering both physically and mentally.[8] *Norwood* seeks to normalize this thoroughly abnormal situation by turning the past into a matter of perspective. This is why Tommy Taft,

6 For a discussion of the *Sketch Book* see Lee and Young 2007.

7 Beecher himself was clearly aware of the gruesome reality behind these pictures. In 1865, after spending most of the war in Europe stumping for the Union cause, he traveled to Fort Sumter and was shocked to witness the scope of the destruction and suffering that Sherman's march had brought to the former Confederacy. It was under this impression that he made his speech of consolation before the Southern population at Fort Sumter. See Stout 2006, 422.

8 In *Gunshot Wounds*, an 1864 medical work on the treatment of Civil War veterans, Mitchell wrote that nerve injuries often disturbed the "correspondence between the visible wound and the interior pain," a "hysterical" symptom that was taken as a feigning of roles and that often led to lasting insanity. See Long 2004, 31.

another sentimental father figure and a representative of the less educated village working class, emerges in *Norwood* as a figure for comic relief. Like thousands of Civil War soldiers he has lost one leg. But unlike the majority of these amputees, who were between 18 and 45 years old (see Huddleston 2003), Tommy lost his leg *before* the war. For him "a wooden leg is a good thing" because he never has "to cut [his] toe-nails on that leg." Says Tommy "with a chuckle":

> Not much paid out for shoes neither. Go to a blacksmith for my shoes – ho! Ho! Ho! Never have rheumatism in that leg either. Don't catch cold when I git it wet. [...] Sometimes I think legs were a mistake; ain't worth as much as it costs to keep "em up." (151)

Tommy is a classical 'local yokel' who makes the best of his situation. Sitting on a pork-barrel, eating hickory nuts, he derives his humor and vitality from his upbringing in rural America. According to the novel, his way of dealing with what for others is a traumatic experience is to apply the stamina and resilience best acquired in the healthy atmosphere of countryside living. A typical sidekick figure and stock character of rural comedy, Tommy was reportedly a favorite among the readership of *Norwood;* his jolly reaction to disfigurement was not seen as a disproportionate comment on that atmosphere of despair and mourning that still preoccupied American society. His message, in other words, hit a nerve, and is central to the novel. Time and again *Norwood* insists on the absolute power of the mind:

> Sickness is very largely the want of will. Everything is brain. There is thought and feeling not only, but will: and will includes in it far more than mental philosophers think. It acts universally, now as open mind, and then just as much upon the body. It is another name for life-force. Men in whom this life or will-power is great, resist disease, and combat it when attacked. (29)

Borrowing from the military jargon war veterans were well acquainted with, the passage targets the veterans among the novel's readers. *Norwood* thereby *masculinizes* the act of reading itself: by addressing a male audience, and suggesting that what today is known as trauma can be overcome by an act of will, the novel makes *imagination* key to individual change and male reinvigoration. Beecher's contemporaries seem to have understood this message very well. As the unnamed reviewer in an 1867 issue of the *New York Times* wrote enthusiastically, *Norwood* is a 'reading trip' with the power to heal a needy population:

> [T]he careless flow of [the novel's] style is like the free discursiveness of its thoughts. [...] [I]t is luminous and living throughout with kindly and noble feeling, and with the contagious cheerfulness of a happy nature; it is a thoroughly healthy and healthful book [...]. (Anonymous 1867, 5)

Beecher's effort to strictly separate the immediate past from the present, and the way he spatializes experience as biologically determined and group-specific, is important aspects of the book's 'healthy' dimension: in the new republican garden there is a place for everyone and everything. If there are aspects that do not fit into its 'natural' pattern there are alternative realms and spaces to resort to that will keep the garden itself in its ideal pastoral shape.

"Gettysburg" is a case in point. Beecher cannot be accused of ignoring the war *tout court*, as he in fact reserves an entire eponymous chapter for the experience. Significantly, however, the event itself is symbolically neutralized. The novel accomplishes this by symbolically feminizing the war, thereby catering to younger, female readers who made up an important part of that varied audience that Beecher needed to address if he was to secure his role as spokesman for middle-class America. In the "Gettysburg" chapter we are introduced to Rose, the novel's female protagonist whose floral name was the inspiration for the rambler rose pattern that decorated the cover of the 1867 edition of the novel. During the war Rose tends wounded soldiers and—contrary to a woman's historical options at the time—even performs surgery. Rose, in other words, is a superior healer who during times of national crisis becomes God's deputy on earth. Yet her power to heal and soothe also has strong symbolic dimensions: focalized through this morally pure and intellectually strong mixture between a true and a new woman, the wartime landscape of destruction becomes a meaningful, sublime battlefield in America's holy war. In the course of this spiritual compensation, Gettysburg takes on the proportion of a nineteenth-century panoramic painting. Never losing visual control over the scene that unfolds before her, Rose "obtains a free view" of the battle as it is carefully framed by "the blue mass of the South Mountains banked up against the horizon" (473). After encapsulating Gettysburg as part of a landscape that is 'bigger than life,' the novel lets Rose find her *real* destination in the concrete, local environment of Norwood. When the war is over, this angelic woman—who has been introduced as the village doctor's exceptionally intelligent, highly educated daughter—happily embraces the role of a wife and mother: shaped by the changing needs of her respective environments, she follows the call of a woman's nature.

Rose's carefree, unregretful return to a conservative model of femininity is just one instance of the novel's investment in the concrete political issues of the postwar period. Published during the tumultuous phase of Constitutional Reconstruction (1865–1869), *Norwood* thrives on earlier republican ideals, Christian values, and a transcendentalist-inflected belief in organic unity, but celebrates the postwar nation's mix of social classes, ethnicities, and racial backgrounds. Staged as the cradle of the new, democratic nation, the ethnically and racially mixed New England village with its educated elite and highly esteemed, down-to-earth common men (Tommy Taft is one of them) and women is cast as model for an ideal social organization to come.

One member of this celebrated multicultural mix is Pete Sawmill, a man whose name signals a worker's 'natural' destination. Supervised by Rose and her benevolent father, he is the symbolical heir of the antebellum house slave who "good-naturedly, and as a matter indisputable, […] [does] as he [is] ordered." In marked contrast to the scholarly whites of Norwood, Pete is excessively physical and can "shoulder his [master's] horses" if told to do so (128–129). Interestingly, this postwar primitive is a defining member of the harmonious little community of Norwood: during the Civil War, readers are told, Pete saved the future village leader, Barton, from certain death in a Confederate military prison. Driven by "an instinct derived from his pre-existent state," the black soldier, Pete, lifted an unconscious Barton "off the ground with his teeth" and carried him safely across a "wild river" (511). This ape-like creature arrives at Norwood with the "gurgles and laughing" of a witless child. By representing the black soldier as absolute, primitive other, Beecher supports the efforts of white Veteran groups to ignore and marginalize the historical contributions of African American soldiers to the war effort. Resembling an animal from that threatening wilderness that is the defining other of the pastoral design, Pete receives a maximum of recognition from an old white woman who exclaims: "Why, you poor old soul, […] we are all as glad to see you, Pete, as if you belonged to us." Pete immediately embraces this crude notion of belonging and identifies himself as a voluntary slave: "He, he, he! I guess I do. I don't b'long to nobody else, except the gin'ral" (472). In a maneuver of imaginary space-making that is typical for the novel as a whole, he eventually becomes the foreman of a group of freedmen who work as farmhands for a Quaker family. The novel, in other words, reconstructs a racialized division of work and life in many ways reminiscent of that Southern plantation life which six years earlier had led to insurmountable tensions between the states, representing it as postbellum ideal for the nation at large. The role of the Northern 'negro', Pete, as accepted leader among his as yet 'untamed' Southern brethren makes slavery's methods of forced labor and corporeal punishment obsolete: in the harmonious community of Norwood *everybody* finds his or her 'natural' place—including the ex-slaves who stand "at a little distance" from the celebrating Norwoodites, "grinning and looking at Rose in the most extraordinary manner" (466). It is only logical, then, that in the end a fence separates the black community from the new generation of whites who eagerly plan the future of their village: in keeping with David Blight's central argument in *Race and Reunion*, the country's African American population pays the prize for white sectional reconciliation. Contrary to antebellum novels like *Northwood* and *Uncle Tom's Cabin*, Pete and his group of farmhands are urgently needed to secure the pastoral life of the leisurely, enlightened white population: kept in a designated space behind a fence, they belong to a man-made, 'tamed' landscape whose sharply distinguished zones are the result of a capitalist, and racist, division of labor.

Importantly, however, the individuals who make up Norwood—including the ex-slaves—are entirely self-reliant, as they act in accordance with their own natures. Because they follow the predilections of their gendered or raced selves, they are independent of the forceful mechanisms of individual self-regulation and communal responsibility that had been a decisive part of the eighteen-century republican contract: in *Norwood,* the *new* republican garden effortlessly regulates itself, creating a natural, Darwinian pattern of inequality that is based on the enlightened insight of each individual in his or her natural—and God-given—limitations.

As I mentioned earlier, the years following Beecher's Fort Sumter speech had damaged his initial reputation as moral leader and guide. With its lengthy parlor talks and philosophical strolls in the garden, *Norwood* offered him an alternative opportunity to explain his political and religious agenda in great detail without making any concrete political or legal suggestions. Eager to establish himself as spokesman of an ideal America, Beecher bases the effortless harmony of Norwood and the happiness of all its inhabitants on a crude mixture of postwar scientific racism, biological concepts of gender, and republican values, thereby smugly circumnavigating the historically concrete, racial, gendered, economic, and political tensions of the era. By dedicating chapters to the black man's life-saving qualities and the white woman's self-reliant decision to leave her post as Civil War nurse and doctor, he makes sure that there is space for everyone in the literary garden of *Norwood*—within the narrow confines of short chapters, and supervised by the controlling logic of a white, patriarchal worldview.

For all its proverbial "woodenness" (Ward and Waller 2000, n.p.) *Norwood* inspired the very real geography of New England (and elsewhere[9]): trying to attract new inhabitants, the second parish of Dedham, Massachusetts, in 1872 advertised itself as a material version of Beecher's fictional "village upon a hill" and renamed itself after the novel's title (Wood 1991, 41). Yet not all Americans considered Beecher's vision of self-regulating hierarchies desirable. White women from the Northern states in particular were frustrated by the antifeminist backlash of the Reconstruction period, as they were denied the vote while male ex-slaves could—at least theoretically—access the ballot beginning in 1868. This constellation, then, makes it seem worthwhile to end this paper with a brief outlook at *Mizora,* a white American feminist's response to neo-patriarchal ideas like Beecher's.

In 1881, one of the leading American feminists, Elizabeth Cady Stanton, recalls the time around which *Norwood* was first published as an era of severe gendered strife:

> Woman must lead the way to her own enfranchisement, and work out her own salvation with a hopeful courage and determination. [...]. She must not put her trust in man in this transition period, since while regarded as his subject, his inferior, his slave, their interests must be antagonistic. (451)[10]

In 1880 and 1881, Mary Bradley Lane from Ohio seems to have come to a similar conclusion and published the first installments of her *Mizora* in the *Cincinnati Commercial.* The literary utopia, which came out as a book in 1889, challenges Beecher's fantasy of effortless subordination by introducing a radical alternative: a white, matriarchal society of the future. Inhabited exclusively by a tribe of beautiful, broad-chested blondes, this all-

9 There is at least one additional example that this was not the only renaming. See http://www.ridleytownshiphistory.com/norwood_history.htm. 10.07.2014.

10 Stanton's statement was a reaction to the Kansas referendum of 1867 that aimed to remove franchise restrictions on the grounds of both race *and* sex. The referendum failed. See Simon and Danziger 1991, 2.

female arcadia looks pretty much the same as Beecher's New England pleasure fantasy—except for the gender of its inhabitants:

> Far, far as the eye could follow it, stretched the stately splendor of a mighty city. But all the buildings were detached and surrounded by lawns and shade trees, their white marble and gray granite walls gleaming through the green foliage. Upon the laws, directly before us, a number of most beautiful girls had disposed themselves at various occupations. Some were reading, some sketching, and some at various kinds of needlework [...]. A large building of white marble crowned a slight eminence behind them. Its porticos were supported upon the hands of colossal statues of women [...]. I supposed I had arrived at a female seminary [...]. (16)

Like the Norwoodites, the blonde Mizorans are entirely dedicated to perfecting themselves and their country. While the village of Norwood is construed around a library, the central building of Mizora is a college. If in Beecher's novel it is mainly the men who engage in philosophical debate, here it is only women who desire to learn and perfect their perfect little republic. Unlike the Norwoodites, the Mizorans work all the time; and unlike Rose they enjoy not subordination but teaching. While the latter eventually prefers pensive strolls in her well-kept flower garden to her successes as a female surgeon, they eagerly conduct experiments in one of the many laboratories; or they take care of other women's daughters with professional skill. But who, then, does the heavy work? Who plows the fields? Cleans the house? An unskilled, dependent workforce does not exist, and there are no slaves in this society of radically similar females: these women have done away with race as a step in self-purification and self-authorization as founders of their own nation. Men, too, are mysteriously lacking: as the novel reveals, men have become extinct in the course of a Civil War and its chaotic aftermath. In the book's revisionist version of Reconstruction, the surviving women managed to sever human reproduction from the heterosexual act. Additionally, they have managed to replace nature and automate work, thereby reconciling the pastoral ideal with the late nineteenth-century technological revolution:

> [T]he products of the orchards and gardens surpassed description. Bread came from the laboratory, and not from the soil by the sweat of the brow. Toil was unknown; the toil that we know, menial, degrading and harassing. Science had been the magician that had done away all that. (21)

Contrary to *Norwood,* where machines are suspiciously absent from the bucolic fantasy, technology is the rallying cry of this feminist success story where machines take the place of 'the negro' and all men, reworking the debris of 1865 into a republican garden. Bringing that celebration of technology that according to Leo Marx was a defining aspect of nineteenth-century United States popular culture at large (see ibid., 375) to its radical, logical end, the Mizorans live in a world that is entirely artificial (even the apples are chemically produced), and have thereby found an answer to all the latent problems that Beecher's nostalgic dream of natural (and ultimately technology-wary) self-regulation was unable to solve. By "eliminating" both the men and the "dark races" through artificial selection and by reproducing chemically (92–93), the blondes have created a radically homogenous society that relies on technology to guarantee the harmonious beauty and peace of their superior lifestyle. While in *Norwood* the pastoral remains tied to a raced and gendered division of labor, the park-like country of Mizora is the work of machines that have replaced earlier definitions of unskilled labor as a woman's or a person of color's 'natural' occupation. Importantly however, neither *Norwood* nor *Mizora* describe their leisurely, education-hungry societies as the result of brute revolutionary force: both the African-American farmhands in Beecher's postwar fantasy, and the machines that make the Mizorans' lifestyle possible are the logical end result of 'natural' selection. In *Norwood* it is the 'natural' superiority of New England whites that authorizes their role as national 'gardeners'; in *Mizora* it is the equally inborn qualities of white women that guarantee the survival of their 'race' and allow them to develop an essentially posthuman society. And yet the Mizoran fantasy of genocide, social hygiene, and an army of perfect women is even more absolutist and antidemocratic than Beecher's gated landscape. Based in what Asha Nadkarni has termed "eugenic feminism"—a pseudo-Darwinian concept aimed at female self-purification for the purpose of furthering the national whole (Nadkarni 2006, 221)[11]—the country of the blondes marks the triumph of (female, white) art over (male, 'black') nature. Barely able to gloss over its racist and anti-male implications, this utopian concept of the garden is essentially posthuman. Fortunately, no American developer chose the name "Mizora" for a new suburban settlement.

11 As Nadkari has pointed out in her discussion of feminism in the US and India, there is an uncanny connection between feminism and racism: "I use the phrase 'eugenic feminism' to refer not only to U.S. and Indian feminism's historical engagement with the eugenics movement but also to the rhetoric of feminism itself. Eugenic feminism is a self-purifying and self-perfecting rhetoric that works to create a feminist subject who, free of race, guarantees the reproduction of the sovereign nation." (2006, 221).

Words Cited

Anonymous (1867). Review of Norwood. *The New York Times*, August 7, 1867, 5.

Applegate, Debby (2007). *The Most Famous Man in America: The Biography of Henry Ward Beecher.* New York: Three Leaves Press.

Beecher, Henry Ward (2010). *Norwood: Or, Village Life in New England.* New York: Charles Scribner.

— (1859). *Plain and pleasant talk about fruits, flowers and farming.* New York: Derby & Jackson.

— (1844). *Seven Lectures to Young Men on Various Important Subjects.* Indianapolis: Cutler. 23.09.2013 http://archive.org/stream/sevenlecturestoy00beec#page/-n1/mode/2up.

Blight, David W. (2001). *Race and Reunion: The Civil War in American Memory.* Cambridge: Harvard University Press.

Butler, Leslie (2006). Reconstructions in Intellectual and Cultural Life. In: Thomas J. Brown (ed.). *Reconstructions: New Perspectives on the Postbellum United States*, 172–205. New York: Oxford University Press.

Clark, Clifford E. (1978). *Henry Ward Beecher: Spokesman for a Middle-Class America.* Chicago: University of Illinois Press.

Daly, Augustin (1867). *'A Legend of 'Norwood,' or, Village Life in New England: An Original Dramatic Comedy of American Life, In Four Acts, Founded on a Novel by Rev. Henry Ward Beecher.* New York: privately published.

Emerson, Ralph Waldo (1965). *Selected Essays, Lectures, and Poems of Ralph Waldo Emerson.* Ed. Robert E. Spiller. New York: Washington Square Press.

Faust, Drew Gilpin (2008). *This Republic of Suffering. Death and the American Civil War.* New York: Random House.

Gardner, Alexander (1959 [1866]). *Alexander Gardner's Sketchbook of the War.* Mineola: Dover Publications.

Grant, Susan-Mary (2000). *North Over South: Northern Nationalism and American Identity in the Antebellum Era.* Lawrence: University Press of Kansas.

Hale, Sarah Josepha (1827). *Northwood: A Tale of New England.* Boston: Bowles & Dearborne.

Hounshell, David (1985). *From the American System to Mass Production, 1800–1932.* Baltimore: Johns Hopkins University Press.

Howells, William Dean (1867). Famous Americans of Recent Times. *The Atlantic Monthly*, May 1867, 636–638.

Huddleston, John (2003). *Killing Ground. The Civil War and the Changing American Landscape.* Baltimore: Johns Hopkins University Press.

Kasson, John F. (1977). *Civilizing the Machine. Technology and Republican Values in America, 1776–1900.* New York: Penguin.

Lane, Mary E. Bradley (2000) *Mizora: A Prophecy: An 1880s Radical Feminist Utopia.* Ed. Jean Pfaelzer. New York: Syracuse University Press.

Lee, Anthony W., and Elizabeth Young (2007). *On Alexander Gardner's Photographic Sketch Book of the Civil War.* Berkeley: University of California Press.

Long, Lisa A. (2004) *Rehabilitating Bodies: Health, History, and The American Civil War.* Philadelphia: University of Pennsylvania Press.

Marx, Leo (2000 [1964]). *The Machine in the Garden. Technology and the Pastoral Ideal in America.* New York: Oxford University Press.

McLoughlin, Thomas (1970). *The Meaning of Henry Ward Beecher: An Essay on the Shifting Values of Mid-Victorian America, 1840–1870.* New York: Knopf.

Nadkarni, Asha (2006). Eugenic Feminism: Asian Reproduction in the U.S. National Imaginary. *NOVEL: A Forum on Fiction*, 39.2, 221–244.

Quigley, David (2004). *The Second Founding: New York City, Reconstruction, and the Making of American Democracy.* New York: Hill and Wang.

Simon, Rita J., and Gloria Danziger (1991). *Women's Movements in America: Their Successes, Disappointments, and Aspirations.* New York: Praeger.

Smith, Henry Nash (1974). The Scribbling Women and the Cosmic Success Story. *Critical Inquiry*, 1.1, 47–70.

Stanton, Elizabeth Cady, Susan B. Anthony, and Matilda Joslyn Gage (eds.). (1881). *The History of Woman Suffrage.* 2 Volumes. New York: Fowler & Wells.

Stout, Harry S. (2006). *Upon the Altar of the Nation: A Moral History of the Civil War.* New York: Penguin.

Ward, Adolphus W., and Alfred R. Waller (2000). Henry Ward Beecher. In Ward and William P. Trent, et al. *The Cambridge History of English and American Literature: An Encyclopedia in Eighteen Volumes.* Volume XVI. New York: G.P. Putnam's Sons, 1907–1921; New York: Bartleby.com, 2000. 21.07.2014 http://www.-bartleby.com/226/1313.html.

Wood, Joseph S. (1991). "Build, Therefore, Your Own World": The New England Village as Settlement Ideal. *Annals of the Association of American Geographers*, 81.1, March 1991, 32–50.

The Dinosaur in the Garden: From Walden Pond to *Jurassic Park*[1]

Susanne Hamscha

Introduction

"The pastoral ideal has been used to define the meaning of America ever since the age of discovery, and it has not yet lost its hold upon the native imagination," Leo Marx writes at the outset of *The Machine in the Garden* (1964, 3). Very clearly, the purpose of his book was to find a usable past[2] through the study of literature, and the pastoral ideal with its promise for a new beginning in an unspoiled environment only seemed a fit theme for this endeavor. In an early review of *The Machine in the Garden*, Charles L. Sanford notes that Marx's book contributed to the understanding of the "cultural and intellectual history of America" (1965, 272) and successfully "reinterpreted America for the atomic age" (ibid. 274). Following Matthew Arnold's basic premise that a national literature reflects its society, Marx turns to "the American experience" as a basis for his analysis (1964, 4). Written against the backdrop of the Cold War, McCarthyism, the failure of the old American left, and the crisis of the 'liberal tradition,' *The Machine in the Garden* is an attempt to delineate the 'essence' of Americanness through a recourse to the nation's literary and cultural past.

While the premise of Marx's study is certainly problematic and needs to be viewed critically for its promotion of a rhetoric of American exception-

1 A slightly different version of this essay has been published as a chapter in my monograph *The Fiction of America. Performance and the Cultural Imaginary in Literature and Film* (2013).

2 The notion of a 'usable past' was introduced by Van Wyck Brooks in his 1918 essay "On Creating a Usable Past," in which he lamented the lack of a cultural memory that could provide American writers with a sense of continuity and a rich literary tradition. As such a cultural memory was not based on any objective 'reality' in the first place but rather a highly subjective, self-styled project, Brooks suggested that the creation of a usable past could correct that apparent failure and provide America with its own national heritage in the arts.

alism, its influence on later generations of Americanists cannot be denied. In this paper, I want to bring Leo Marx into dialogue with Paul Lauter, whose essay collection *From Walden Pond to Jurassic Park* (2001), written on the brink of 9/11 and the ensuing war on terror, is in many ways a reply to *The Machine in the Garden*. Lauter ends his study with a lengthy quotation from Marx's study, but it is particularly the underlying tension between politics and aesthetics, or art, which connects their work—even though it goes without saying that they come to entirely different conclusions. In his chapter "Dinosaur Culture" Lauter engages most obviously, albeit not explicitly, with Marx's work. Lauter's concise analysis of *Jurassic Park*'s negotiation of the boundaries between nature, culture, and technology in that chapter suggests that one can read Spielberg's film as a re-articulation of the American pastoral in a post-industrial, neoliberal age. Lauter's chapter reads very much like a reply to Marx's discussion of the pastoral ideal in Henry David Thoreau's *Walden* adapted to the conditions of the late twentieth century. Both *Walden* and *Jurassic Park* are meditations on the contemporary state of American culture, and both texts express the yearning for a simpler, purer, and more immediate experience of the self in its environment. While I do not want to fall back into a rhetoric of American exceptionalism and suggest that there is something essentially 'American' about *Walden* and *Jurassic Park*, I do want to make the point that both texts are intimately tied to questions concerning the meaning of 'America'; finding fault in America's contemporary condition, both texts try to rediscover a purer and better version of 'America' through a nostalgic return to a virgin land that has not yet been corrupted by capitalism, technology, and industrialization.

In *The Machine in the Garden*, Leo Marx reads *Walden* as "the report of an experiment in transcendental pastoralism" (1964, 242); in this paper, I suggest that one can approach *Jurassic Park* as an experiment in technological pastoralism. While the idyllic environs of Thoreau's pond are disturbed by the railroad, the harbinger of a new age, the primordial nature of Isla Nublar, a fictional island just off Costa Rica, has to succumb to the technologically engineered T-Rex in *Jurassic Park*, which symbolizes the intricate relation between nature and technology in a post-industrial society. My paper will identify three key themes which *Walden* and *Jurassic Park* share and argue that Thoreau's biggest nightmare has transformed into the shape of a dinosaur. The juxtaposition of *Jurassic Park* with *Walden* shows that the dinosaur is an intensification of the machine in the garden: it signifies the ultimate outgrowth of capitalism and excessive commodification, and, as a creature born jointly out of natural and technological processes, it complicates Marx's dualism. In *Jurassic Park*, machine and garden neither exist side by side nor is the garden penetrated and disturbed by the machine, but the two have merged and become dependent on one another, one not able to exist without the other.

By bringing *Walden* into dialogue with *Jurassic Park* and Leo Marx into dialogue with Paul Lauter, my paper also reflects the changing face of American Studies from the Cold War to the War on Terror. Marx and Lauter's books are responses to the concerns and agendas of the discipline at two different decisive historical moments in time, and both contemplate the role America will play in such times of great political change and upheaval. Their interpretations of *Walden* and *Jurassic Park* are, therefore, also interesting snap-shots of disciplinary self-reflection, which tell the reader as much about the texts discussed as they do about the contemporary state and the future of American Studies.

From Walden Pond to *Jurassic Park*

According to Leo Marx, Henry David Thoreau's masterpiece *Walden* is a "distinctively American version of the romantic pastoral" (1964, 245). Set somewhere between Massachusetts and a utopian land of fantasy where myth blends with reality, Thoreau's hut stands at the center of a symbolic landscape, on the border between the village of Concord and pure, unaltered nature. Thoreau describes the site of his hermitage as an ideal pasture: Located not even two miles south from the center of Concord, Walden Pond is a real place which is transformed into a fantastic landscape of the mind, unaffected by the constraints of space and time, as Marx reads Thoreau's account. In form and feeling, Marx suggests, *Walden* follows the conventions of the classic Virgilian mode of the pastoral but that affinity is disguised by the peculiarities of American pastoralism in this book (see 1964, 244). Marx cites the literalness with which Thoreau seems to approach the withdrawal from society and ideal of a simpler life as the most striking feature of this particularly American pastoral, even though Thoreau notes several times that he sees himself neither as a hermit nor a social dropout. Rather, the proximity to Concord allows him to lead a life in casual solitude; when he moves into his hut at the pond, he retreats from a social community to which he was obliged to contribute his share as long as he lived in its midst, but at the same time he is free to connect with the people of Concord whenever he pleases.

Thoreau moved into his hut at Walden Pond on Independence Day 1845; "by accident," as he says, but the symbolism of choosing that particular day cannot be denied (1983, 128).[3] The book is set up as a single man's individual Declaration of Independence, his search for ultimate personal, intellectual, and political emancipation, and thus as a fable of the renewal of American culture. As Frederick Garber explains, the experiment of self-discovery which Thoreau articulates in his dual function as author and narrator—the focus on an individual's "self-refashioning" and "remaking of origin within [itself]"—serves as metonymy for the "self-reoriginating" of American society and culture (1991, 192–193). The rediscovery of America is only the effect and not the purpose of Thoreau's experiment, however. As Larzer Ziff points out, Thoreau believed that his writings and the example of his life would serve as a general lesson on how to keep balance between being part of a community and pursuing one's private interests, but ultimately also served as a lesson on how to improve the current state of the American nation. (see 1981, 197). Despite the universal aim of the text, *Walden* can, therefore, be read as a particularly 'American' text, not least because "its facts and images are drawn from the circumstances of life in nineteenth-century America," as Marx notes (1964, 246–247); it is written against the backdrop of a cultural malady which Thoreau diagnoses in the opening chapter "Economy" and grapples with throughout the text. He notes that his townsmen in Concord have become "serfs of the soil," men "crushed and smothered under [the] load" of their daily labor, which keeps them from experiencing and enjoying much of

3 Thoreau's decision to move to Walden Pond on Independence Day is peculiar—after all, he "did not think much of America," as he told Walt Whitman (1958, 445). Maybe he saw his experiment as a reenactment of the first Puritan settlements rather than a second Declaration of Independence, and therefore found the Fourth of July the wrong day to take up the abode (see Cavell 1992, 8). Bob Pepperman Taylor has argued that "[t]he Revolution, with its heroes, principles, noble deeds, and ideals, is the story we like to tell of our founding. But Thoreau floats by the remains of this event, and [...] return[s] to the theme of settlement of New England by Europeans" (1996, 20). Thoreau, Taylor writes, wanted to discover earlier times, "when the fate of the new civilization was still unsettled, when white settlements were themselves new, untested, and unsure of their futures" (1996, 20).

what life has to offer (1983, 47). Most men do not reflect on the work they do, nor question the value and necessity of it, but they blindly obey to convention, on the one hand, and to a new spirit that is rampant in mid-nineteenth-century America: capitalism.

Thoreau is aware of the fact that capitalism is not merely an economic system but a new culture and mode of perception that renegotiates values and meanings on a big scale. Money has always been the distinguishing marker between rich and poor, but industrialization and technological progress have transformed the economic and social landscape in unprecedented ways. The accumulation of capital, the fierce competition on the market, and wage labor create new forms of dependencies and limit the individual in its freedom:

> Actually, the laboring man has not leisure for a true integrity day by day; he cannot afford to sustain the manliest relations to men; his labor would be depreciated in the market. He has no time to be anything but a machine. (1983, 48)

The impression that growing industrialization has usurped man to such a degree that he has become a technologized being is disconcerting to Thoreau. A look at his Concord neighbors only shows him that they "have been toiling twenty, thirty, or forty years, that they may become the real owners of their farms, which commonly they have inherited with encumbrances, or else bought with hired money [...] but commonly they have not paid for them yet" (1983, 75). Caught in a web of interdependencies determined by monetary relations, Thoreau's neighbors are constantly in danger of breaking down morally, of distancing themselves from nature, and of putting convenience and commercial interest above their individuality and their personal independence. Thoreau senses that man has been co-opted by the new economic system and turned into the extension of the machines he has created to increase productivity. In taking up his abode at Walden Pond and leading a 'self-sufficient' life in harmony with nature, Thoreau confronts these new developments by distancing himself from society, thus reclaiming his individual freedom and independence.

Nature is the only sphere that has "a kind of literal authority precisely because she is not one of men's institutions," as Walter Benn Michaels suggests (1977, 138). It has the capacity to give life a deeper meaning, as Thoreau notes, happy that, unlike the days of his fellow townsmen in Concord, his days in the woods are not "bearing the stamp of any heathen deity," that is, that of money (1983, 157). In experiencing nature, he can shake off the toils of modernity, industrialization, and rising capitalism, which keep man in chains. In the towns and villages, Thoreau observes, man is his own slave-driver, but in its idealized, pastoral form nature presents man with freedom, emancipation, and a newfound appreciation of the essential facts of life. Society has been dazzled by the advances of a materialistic, industrial era which has transformed men into commodities and reduced them in their divinity. At Walden Pond, Thoreau turns back the clock and enters a world that seems to be free of standardization and competition in the marketplace, free of human exploitation and objectification, free of conformity and submission to institutions. He finds amusement and entertainment in observing nature and studying the animals in the woods, in listening to nature's sounds and learning to read its signs. A wide range of common activities, such as reading, fishing, watching the sunset, observing birds, hoeing beans, and laying bricks become charged with possibility and unexplored potential as Thoreau performs them. These activities connect Thoreau with the idyllic nature he creates in his text and teach *Walden's* readers to find "the great, unsuspected possibilities in ordinary life" (Cafaro 2002, 23). The inconspicuous shack in the woods becomes the symbol of Thoreau's resistance to the materialistic and economic expectations of the capitalist culture he lives in, the pastoral ideal that allows him to measure time, gain, and success by the counts of nature rather than the counts of money.

Jurassic Park, I want to suggest, picks up many of Thoreau's concerns and demonstrates what the outgrowth of capitalism and commodification can lead to. Steven Spielberg's enormously successful 1993 blockbuster spins a tale of the desire to escape from a world pervaded by technology, mass production, and consumerism and immerse oneself in primordial nature. Paradoxically enough, however, the natural world in *Jurassic Park* has been created through the very means that the protagonists seek to escape from. The film is set on the fictional island Isla Nublar, where billionaire and philanthropist John Hammond (Richard Attenborough) and his team of scientists have built a nature theme park of cloned dinosaurs. Threatened with legal action over the accidental death of a park employee, Hammond invites paleontologist Alan Grant (Sam Neill), paleobotanist Ellie Sattler (Laura Dern), and chaos theorist Ian Malcolm (Jeff Goldblum) to inspect the park before its official opening and confirm that all measures have been taken to ensure the public's safety. Hammond's two grandchildren and a corporate lawyer join the three scientists on their guided tour through the park. Along the way, their electric tour cars break down due to sabotage on the park's security system engineered by a rival company. As a tropical storm is fast approaching the island, most park employees leave the island on a boat for the main land, but with his guests stranded in the park, Hammond and some of the staff stay on the island to reboot the electric system and regain control over the park. The dinosaurs, including carnivorous raptors, have already been set loose, however, and Hammond's visitors and team need to escape the island under peril before it is too late. In the end, Hammond, Alan, Ellie, Ian, and the children are rescued by a helicopter after the storm has settled, while the park's remaining employees are left behind and presumably killed by the dinosaurs.

While *Jurassic Park* certainly is a typical action movie that privileges suspense and special effects over complex characters and a sophisticated plot, it captures a familiar problem—a Thoreauvian problem, one might say—on a deeper level. *Jurassic Park* is a film about economy and morality, about the proliferation of technology, and about the attempt to restore a long lost past and immerse oneself in a seemingly primordial, untamed nature. In that sense both *Walden* and *Jurassic Park* use the pastoral mode to narrate the clash between the idea of harmonious accommodation to nature and the progressive quest for wealth. To be sure, the pastoral episodes in *Jurassic Park* are very brief and almost instantly displaced by the gruesome and horrifying ways in which the technologically infused natural world causes damage to its creators. As I argue, *Jurassic Park* can be read as an extrapolation of *Walden*, as it picks up on Thoreau's thoughts on economy, capitalism, and nature, which it adapts to the socio-economic context of the 1990s to delineate the trajectory of the concerns Thoreau voiced almost 150 years earlier.

In contrast to Walden Pond, *Jurassic Park*'s pastoral landscape is notably located outside US territory, on an island just off the Costa Rican coastline. In this foreign and exotic place, one can still encounter primordial nature, which has become impossible within the national boundaries of a hyper-technologized United States. Thoreau only had to leave the town limits of Concord to discover a supposedly purer version of America, but in *Jurassic Park* that imagination of 'America' can only be discovered through its projection onto seemingly uncharted territory.[4] As Anne Brigham suggests,

4 The rediscovery of 'America' elsewhere contains a narrative of imperialism and colonization, in which all signs of indigenous cultures and populations have been erased from the landscape onto which imaginings of 'America' are projected. Mary Louise Pratt has used the term "anti-conquest" to describe the actions of the "seeing-man," the white

Jurassic Park combines "a vulnerability to obsolescence with a fear that we are no longer discoverers, unable to produce anything new or authentic" (2002, par. 10). Just as in *Walden*, the immediate exposure to nature and wilderness is supposed to generate an original and authentic experience, unencumbered by the pressures of a profit-oriented economic system and the coolness of an alienating, clinical social environment. This becomes particularly evident when the film's protagonists, Alan Grant and Ellie Settler, are introduced. Their first appearance in the movie takes place at an excavation site in Montana, where they examine dinosaur skeletons, trying to figure out how those raptors died. It is quickly made clear in this scene that untamed nature has ceased to exist on US territory. The dinosaur skeleton serves as a metaphor for the loss of wilderness and for man's alienation from nature due to the effects of capitalism, consumerism, and the proliferation of technology—all the factors which Thoreau had already identified as the maladies of modern societies. Nature has become a subject of science, something to be studied and analyzed rather than experienced; Alan and Ellie are portrayed as external to, or above nature, carrying out all their actions in an orderly, controlled, and structured manner, aided by technology and strictly following protocol.

The excavation-scene in Montana is powerfully contrasted with the scene of Alan and Ellie's arrival on Isla Nublar. The camera follows their helicopter as it makes its way through the island's lush greens to an astounding waterfall, where Hammond's helicopter landing platform is the only visible sign of civilization. Hammond, Alan, and Ellie immediately board two jeeps together with Ian Malcolm and the lawyer of Hammond's investors to take a ride through the park. The iron security gate they have to pass seems to be the only reminder that this park is man-made and that the 'wilderness' they encounter is facilitated and controlled by men and machines. As they enter the park, Alan and Ellie are in awe of the nature that surrounds them: a seemingly very pure, primordial nature has been restored in the park, where extinct animals have been brought back to life

---

male subject, "whose imperial eyes passively look out and possess" (2008, 9). Through anti-conquest, she argues, the colonizing subjects "secure their innocence in the same moment as they assert [...] hegemony" (ibid). The American 'colonizers' in *Jurassic Park* see the ostensibly untouched landscape around them and lay claim to it through the fantasies and imaginings this 'paradisiacal garden' triggers. Similarly, Thoreau's act of settlement at Walden Pond "is very much in tune with the imperial fantasy," Marek Paryz writes, because it constitutes a symbolic act of claiming authority over the land (2012, 111).

and extinct plants flourish again. In utter disbelief that what they see is indeed true, Alan and Ellie get out of the jeep to get a closer look at the brachiosaur crossing their way. This is a significant moment, as it symbolizes Alan and Ellie's immersion in nature; they deliberately leave the safe environment of the car, which creates an ostensibly unnatural barrier between them and the wilderness, desiring to *experience* nature rather than merely *study* it.

The problem in the park is of course that the nature Alan and Ellie encounter is not 'authentic' but man-made, and not there solely for pleasure and education—or for its own sake—but also to generate profit. As the film's story is situated in the environs of Costa Rica and as it "eliminate[s] the subjects identified with that space" to follow the adventures of white, upper/middle-class Americans, *Jurassic Park* takes us "behind the scenes of Western capitalist production" and American imperialism (Brigham 2002, par.41). The clear segregation between capitalist production located in the 'third world' and its consumption by the 'first world,' which the film's set-up implies, already breaks down in the opening sequence. In the process of unloading a raptor and putting it into its new enclosure, a park employee, a middle-aged Hispanic man, is accidentally killed and devoured by the dinosaur. As a technologically engineered species, the film's dinosaurs represent the result of a "disembodying proliferation of electronic technologies" which is threatened by the return of a pre-modern, pre-industrial world (Bukatman 1993, 62). Without modern technology and scientific innovation the dinosaurs would not exist, yet their natural drives and instincts let them turn against their creators, which inevitably leads to the implosion of the seemingly clear order that had been set up in the theme park. *Jurassic Park*'s attempt to recover an idyllic state of nature in a world penetrated by technology and following the rules of capitalism produces an interesting spin to the machine-in-the-garden motif, for here the garden, which has been completely annexed by the machine, revolts and turns precisely against that power which sustains it. The following three sections, in which I roughly follow the list of themes Paul Lauter distills from *Jurassic Park*, will illuminate the film's twist on the machine-in-the-garden motif and will further explore its relation to Thoreau's *Walden.* These themes include first, technology getting out of control; second, the monsters getting loose; and third, the ghost in the machine.

Technology out of Control, or: When the Train Becomes a T-Rex

Jurassic Park and *Walden* share a timely problem which drives both texts, albeit in different degrees of intensity, namely mankind's limited capacity to control its own technological innovations and to confront scientific progress with its ethical standards in check. The opening sequence of *Jurassic Park* already indicates that the film's most pronounced theme is that of science and technology out of control. The film suggests that the dangers of science and technology are a consequence of man's hubris and the false conviction that nature can be controlled by technology. However, in Hammond's park, 'nature' and 'wilderness' are highly paradoxical concepts to begin with. From the very beginning it is made clear that the park's nature has its clear physical boundaries, which are demarcated by iron gates and electric fences, and could not exist without the aids of modern technology. Moreover, everything that lives and grows within this fenced-in territory has been created by human hands and shaped by technological equipment. Hammond's dinosaurs were cloned out of DNA extracts from dinosaur blood preserved within mosquitoes fossilized in amber and bred in a laboratory. All dinosaurs are designed to be females, so as to be able to completely control the population of the park, but their strands of frog DNA, which are used to fill the gaps in the dinosaur DNA, enable the dinosaurs to change their gender and procreate.[5] In other words, nature in Hammond's park is hardly untamed, but it proves to be untamable, which embeds the film's central question regarding the controllability of nature in

5 Quite a lot of critical attention has been devoted to the dinosaurs' gender and to the film's reproduction of stereotypical associations between femininity, monstrosity, and consumption. As Brigham, for instance, observes, "these monstrous females represent consumption" and adopt "practices of insatiable devouring and reproduction in the wild that thwart the goals of capitalism and patriarchal order" (2002, par.11). Laurie Briggs and Jodi I. Kelber-Kaye analyze *Jurassic Park* as a cultural depiction of debates about genetic engineering and argue that the film takes an essentialist and conservative anti-feminist stance, as genetic reproduction is represented as "unnatural" and dangerous. *Jurassic Park*, they suggest, calls for a return to natural modes of mothering, thus valorizing the traditional nuclear family as the only "good" and "safe" form of reproduction (see 2000, 94–95). Lisa DeTora suggests that, interestingly enough, "the insertion of the nuclear family into the experiential milieu of genetics [is] the original problem that creates monsters," as the park is explicitly designed as a family-friendly amusement park (2008, 14). As the film progresses, she argues, it becomes clear that the nuclear family cannot save the park but must rather be saved from it.

broader debates on ethics, morality, and the intricacies between science and capitalism. With *Jurassic Park*, Michael Crichton, who wrote the novel the film is based on and co-wrote the screenplay together with David Koepp, explicitly wanted to warn about the dangers "of the commercialization of genetic engineering" (Shay/Duncan 1993, 4), which is why he decided to set the story in a theme park, a place that is by definition designed in the interest of commercialism. As Crichton explains, "[t]he fact that these dinosaurs are made for a park, it seemed to me, emphasized rather nicely the idea that all this amazing technology is being used for essentially commercial and frivolous purposes" (Shay/Duncan 1993, 4).[6]

Jurassic Park suggests that, in a capitalist system, virtually *anything* can be turned into a commodity, which inevitably breeds immorality, arrogance, and disaster. This notion is embodied by two minor characters in the film, the greedy corporate lawyer and the filthy computer nerd who is paid by Hammond's rivals to sabotage the park. Both of them suffer violent and humiliating deaths, which are at the same time grotesquely hilarious spectacles: the former is devoured by a T-Rex as he hides in a bathroom, the latter is eaten in his own car, amidst junk food and garbage, by a cunning dilophosaur. Their deaths bear a certain irony, as they are con-

6 As the movie *Jurassic Park* is itself caught up in an elaborate merchandise machinery, its criticism of capitalism and commercialism appears somewhat ironic. Screenwriter David Koepp reflects on this paradox: "Here I was writing about those greedy people who are creating a fabulous theme park just so they can exploit all these dinosaurs and make silly little films and sell stupid plastic plates and things. And I'm writing for a company that's eventually going to put this in their theme parks and make these silly little films and sell stupid plastic plates. I was really chasing my tail there for a while trying to find out what was virtuous in this whole scenario—and eventually gave up" (Shay/Duncan 1993, 56). In his analysis of the film, Andrew Gordon also notes that *Jurassic Park*'s critique of the "exploitation of technology in the interest of commercial greed" is an "uncomfortably self-reflexive and hypocritical" one (2008, 209), while Constance Balides argues that the movie's presentation of merchandise displays the "lustre of capital itself" (2000, 160). *Jurassic Park* "had an unprecedented marketing campaign with more than a thousand products being licensed officially," ranging from T-shirts to sippy cups, which all featured the by now iconic logo of a dinosaur skeleton (Taylor 1992, 141). Nowhere else is the compliancy of the film with commercialism more obvious than at the Universal Studios theme park in Hollywood, where a "*Jurassic Park* Ride" was opened in 1996, shortly before the release of *Jurassic Park II: The Lost World*. The ride ends right in front of the gift shop filled with *Jurassic Park* merchandise, through which all visitors must exit. When *The Lost World* was released in 1997, Spielberg self-mockingly commented on the huge financial success of the *Jurassic Park* franchise: "I liken myself to the hunters that go after the animals [in *The Lost World*]. They'll do anything for the money, and so will we" (Biskind 2000 201).

sumed in a theme park, that is, in a landscape of capitalist consumption, by the very creatures that seemed to promise them a life of riches. "We can charge anything we want," the lawyer exclaims when he first sees the dinosaurs, "and people will pay it!" Hammond replies that he does not want to overcharge, because he wants his park to be accessible to everyone, but it is made very clear that Hammond's noble wish to restore nature and recover the past is ultimately subject to and dependent upon the structures of capitalism. In other words, both the creation of the park and its execution are permeated by commercial interest, the pleasures of consumption, and the principles of profit.

In *Walden*, the dangers modern society is faced with are at first sight far less menacing, but to Thoreau they are, nevertheless, unsettling. In the chapter "Sounds," for instance, Thoreau's efforts to escape capitalism and industrialization are intermittently interrupted by the "rattle of railroad cars," which deliver goods to the neighboring village (1983, 160). It is the railroad that keeps Thoreau "related to society," as he puts it, but it also serves as his reminder of the dominant position that technology and capital have started to enjoy in his culture (1983, 160). The whistle of the locomotive informs him of the "restless city merchants" and "adventurous country traders" arriving, of the groceries and portions being delivered and paid for, and of the cotton, cloth, and books that are being sold to his neighbors (1983, 161). As the train always travels on the same schedule, its sounds function as Thoreau's timekeeper, letting him keep track of how much time has passed and of how he had put that time to use. In the later chapter "Spring," however, Thoreau introduces a different means of measuring time and its passing, which is the progress of the seasons.[7] By listening to and observing the natural world, he keeps time according to a different, much slower beat, which stresses the contrast between the speedy talk and quick thinking practiced in Concord and the introspective speech and philosophical thought Thoreau enjoys at Walden Pond.

7 As Nina Baym has pointed out, *Walden* consists of twenty chapters "neatly paired in oppositions" (2007, 3): reading and sounds, solitude and visitors, the bean-field and the village, the ponds and Baker Farm, higher laws and brute neighbors. The chapters are arranged in seasonal sequence, beginning with his move to Walden Pond in summer and ending with the arrival of spring. This pattern, as Baym explains, "aligned his stay with myths of death and rebirth, allowing him to identify his narrative … with a psychic journey from joy (summer) through depression and despair (winter) into renewal (spring)" (2007, 3).

As the chapter "Sounds" progresses, however, Thoreau seems to perceive the train as less and less disturbing and comes to integrate its penetrating noise into Walden Pond's sound landscape. When he briefly notes "the distant rumbling of wagons over bridges" after a lengthy description of an owl, whose "insane howlings" reminded him of "ghouls and idiots," the train suddenly seems far less menacing and alien than earlier in this chapter (1983, 171). It is almost as if the train had blended into the environment and become an integral part of it—no less 'natural' to Thoreau's ear than the hooting and screeching of an owl. Still, his getting used to the sounds of the train does not keep Thoreau from taking a critical stance towards the changes the railroad brings about. He is fully aware of the railroad's power in restructuring America's culture and economy, and of the historic significance which an extension of the nation's railway system towards the West might have. Calling the train a "traveling demigod" and "cloud-compeller" Thoreau recognizes its power and appeal, one the one hand, but he also sees the danger of technology taking control over man if greed and fatalism get the upper hand (1983, 161–162). As he resignedly notes, the train moves with such velocity and determination that there is no stopping its triumph, so his only modest wish is that men may use the railroad for noble, heroic, and innocent ends. Thoreau's hope is, needless to say, as forlorn as Hammond's intention to clone dinosaurs solely for educational purposes. In *Jurassic Park*, we are confronted with precisely that version of American culture which Thoreau so dreaded, that is, a culture dominated by capital, technology, and egotism. Hammond's attempt to restore a pastoral idyll through the recreation of a time in which man did not yet exist and therefore had no hand in nature represents an attempt to let the Garden enter back into the Machine, but of course the Machine is a constitutive part of this Garden. In *Jurassic Park*, it is not technology that disrupts nature but nature is rather interspersed into technology; nature is created, maintained, and sustained by technological intervention. Put differently, in its depiction of the excesses of a culture driven by commercial interest, *Jurassic Park* affirms Thoreau's anticipation that capitalism is too powerful a system to be employed with second thoughts.

Monsters Getting Loose

Closely related to the theme of technology out of control is that of the monsters getting loose. In *Jurassic Park*, the monsters are, on the one hand, the dinosaurs which go on a rampage and destroy Hammond's park, and, on the other hand, the genetic engineers and investors who created the theme park and its inhabitants. These 'human monsters' are given hardly any screen time so that they remain a largely invisible and anonymous mass of people, whose actions are determined by scientific and economic reasoning rather than human interaction and moral responsibility. In other words, the geneticists and investors function as the impersonal representatives of the capitalist machinery, who are hardly more than a shadowy presence and yet enormously crucial for the ensuing horror in the park. The combination of techno-scientific research and capitalist ideology proves to be the ultimate monster, then, an unstoppable force that feeds on man's hubris and greed. "God creates dinosaurs. God destroys dinosaurs. God creates man. Man destroys God. Man creates dinosaurs," Ian Malcolm aptly summarizes the reckless play with technology and nature that unfolds before him in the park. 'Dinosaurs destroy man' is the next logical step in this sequence, a step that lets the dinosaurs appear not as 'others' but as extensions of ourselves—as animals, Lauter argues, "peculiarly like ourselves" in their monstrosity and insatiability (2001, 104).

In *Walden*, the visible and tangible 'monster' comes in the shape of the railroad, which is no less rampant and fierce than the dinosaurs. "The whistle of the locomotive penetrates my woods summer and winter, sounding like the scream of a hawk sailing over some farmer's yard," Thoreau recalls the violence with which the railroad cuts through the landscape (1983, 160). The noise with which it moves along makes it impossible to ignore its existence; however, Thoreau's likening of the locomotive's whistle to the scream of a hawk also suggests that the railroad is not experienced as an utterly alien force but rather as a man-made extension of nature, a 'creature' of modernity with immeasurable power. This notion becomes particularly clear in Thoreau's description of the railroad as an "iron horse," which makes "the hills echo with its snort like thunder, shaking the earth with his feet, and breathing fire and smoke from his nostrils" (1983, 161). Thoreau's comparisons of the train and its movement to a gigantic, powerful horse, creates a picture of the train as a hybrid between animal and machine, "a race now worthy to inhabit [the earth]" (1983, 161), quite similar to Spielberg's dinosaurs in *Jurassic Park*. And the railroad, too, is a prime example of monstrosity and insatiability: all day long, it "flies over the country, stopping only that his master may rest" (1983, 162), coming and going with such punctuality and precision that now "one well conducted institution regulates the whole country" (1983, 163). In analogy to Ian Malcolm's logical sequence in *Jurassic Park*, the logic in *Walden* is that man creates railroad and railroad destroys man. "To do things 'railroad fashion' is now the by-word," Thoreau notes, "and it is worth the while to be warned so often and so sincerely by any power to get off its track" (1983, 163). Man has constructed his own fate by letting the railroad take command of how towns and cities organize their daily routine, but he realizes that the railroad's capacity to ease trade and transportation is unprecedented. There is a fascination about the railroad Thoreau cannot deny and it is immediately clear to him that this "iron horse" is the symbol of a new era—the symbol of a faster-paced, profit-oriented world, the symbol of unstoppable technological progress, and the symbol of new imperialist aspirations, as the expansion of the railway system to span the entire continent from north to south and east to west only seems to be a question of time.

Similar to the dinosaurs in *Jurassic Park*, then, the invisible monster looming in the background is capitalism, which is able to thrive with the expansion of the railroad system. The railroad gives this new ideology that is sweeping the country a face; it is the machine that visibly and markedly enters the garden, and, as Marx suggests, its image "figures an ambiguity at the heart of *Walden*":

> Man-made power, the machine with its fire, smoke and thunder, is juxtaposed to the waters of Walden, remarkable for their depth and purity and a matchless, indescribable color—now light blue, now green, almost always pellucid. The iron horse moves across the surface of the earth; the pond invites the eye below the surface. The contrast embodies both the hope and the fear aroused by the impending climax of America's encounter with wild nature. (1964, 251)

The image of the railroad rattling along the pond's shore expresses a tension within Thoreau. Although he is aware of the dangers that come with the swift integration of the railroad into daily life and vilifies technology because it distracts his attention from his (presumably more important) concerns regarding nature, the electric atmosphere, the punctuality and precision, and the order and efficiency of the railroad also elate Thoreau. Furthermore, he is impressed by the adventurousness and confidence of the men who operate the train and venture into new territories. The railroad symbolizes change and "progress," it becomes the "agent and type of an irreversible process," that is, "the implacable advance of history," from which there is no escape (Marx 1964, 252).

The dinosaurs in *Jurassic Park* fulfill a similar function as the railroad in *Walden*: they stand for an ambiguity that is at the heart of this film. On the one hand, they represent a primordial state of nature and express a supposedly desirable return to wilderness while, on the other hand, they are also symbols of the excesses of capitalism and a technological 'progress' that is unstoppable. This ambiguity is enveloped by a moral disagreement between Hammond and his guests, who see that aside from the problem of incalculable safety risks for future park visitors, there is the problem of moral responsibility, which Hammond refuses to address. Malcolm voices his concern that the park might get out of control by comparing Hammond's venture to children's play: "You wield it [genetics] like a kid who's found his dad's gun." As James H. Spence has pointed out, Malcolm seems to believe that nature is above man and "*deserves* to be respected" (2008, 100), whereas Hammond thinks that man ought to use his abilities to "give an extinct species a second chance." However, as Malcolm replies, Hammond's recreation of extinct species actually runs against one of the basic principles of Darwinism, and that is the survival of the fittest: "Your scientists were so preoccupied with whether or not they could do that [recreate dinosaurs] that they didn't stop to think if they should," he scolds Hammond and criticizes the "lack of humility before nature" that is displayed in the park. Malcolm's argument is based on the presumption that "there is something good about nature" and that "what is natural is good" (Spence 2008, 100, emphasis in the original). As the extinction of dinosaurs was not caused by man but (according to Malcolm) can be regarded a prime example of natural selection, bringing dinosaurs back to life would be a violent act of tampering with nature. "What you call discovery, I call the rape of the natural world," Malcolm says to Hammond, reiterating Thoreau's description of the railroad as a penetrating force. *Jurassic Park*'s dinosaurs can thus be read as the superlative to *Walden*'s train, creatures that arouse both fascination and fear and that symbolize the irreversible transgression of a moral boundary motivated by the promise of profit and by the overestimation of man's powers.

The Ghost in the Machine

The third theme that *Walden* and *Jurassic Park* share is that of increasing commodification and its negative consequences, which is expressed through the motif of "the ghost in the machine," as Paul Lauter calls it (2001, 106). *Jurassic Park* offers the point of view that commodification permeates all spheres of life and that virtually anything can be turned into a commodity—which is essentially a dangerous thing. All nature in Hammond's Jurassic Park, including the dinosaurs, is reduced to a "consumer spectacle," where science is practiced "according to the principles of profit rather than knowledge" and nature becomes the product of man-made technological intervention (Stern 2004, 355). At the same time, however, the *Jurassic Park* franchise gets away with the very same thing it criticizes: "Crichton and Spielberg accomplished in art what Hammond [...] within the film could not," Harriet Hawkins notes (1995, 44). "They triumphantly created an island inhabited by dinosaurs," she continues, "who behave exactly according to direction and subsequently made a mega fortune selling the recreated dinosaurs to a global audience" (1995, 44). As a cinematic experience, the film is entirely bound up with the commodification of 'nature' and the commercial exploitation of technological and scientific innovation. At the same time, the film's economic narrative strand tacitly points to the exploitation of human labor force in the process of commodification and the imbalance between 'first' and 'third' world.

Paul Lauter reads *Jurassic Park* as offering a "paradigm on colonialism," as the park has been built on land appropriated from its 'native' inhabitants, one of whom becomes the dinosaurs' first victim (2001, 106). As the dinosaurs are the brainchild of rich, greedy investors and immoral, perhaps even megalomaniac geneticists, the death of the park worker makes clear that within the logic of capitalism the maximization of profit counts more than the life of a local employee. Or, to put it more provocatively, as long as only Hispanic park workers are killed by the dinosaurs, the park is kept running, because they are tolerable casualties. It is only when the white American visitors are in peril that Hammond begins to reconsider whether the theme park can really be kept under control. This form of erasure and repression of the 'other' is what Lauter calls "the ghost in the machine," which is not given narrative form but which "reveals most fully what drives people and societies" (2001, 106). In this sense, the dinosaurs figure as "the visible products of technocolonialization," artificially created monsters formed out of "banal stuff" that threaten their creators and consume "the ordinary folk" who helped to make and raise them but who remain a faceless mass, invisible in the complex chain of production and commercialization (Lauter 107). There is a systematic crisis with regard to consumption on Isla Nublar, as the consumers are (in danger of being) consumed and the socio-economic order is turned topsy-turvy. When their project of colonization fails miserably, the white Americans leave the park the same way they have entered it: with a chopper, "that familiar tool of third world repression" (ibid., 107). The fate of those left behind, the ethnic park workers, genetic engineers, or native peoples on the island remains unknown and is rendered insignificant.

Similarly, Thoreau offers us a glimpse at the effects that commodification has on the Irish workers exploited in the construction of the railroad and on the Native American population in Concord. In "Economy," Thoreau recalls an encounter with a Native American in Concord who tried to sell baskets to the wealthy townsfolk:

> Not long since, a strolling Indian went to sell baskets at the house of a well-known lawyer in my neighborhood. "Do you wish to buy any baskets?" he asked. "No, we do not want any," was the reply. "What!" exclaimed the Indian as he went out the gate, "do you mean to starve us?" Having seen the industrious white neighbors so well off [...] he had said to himself; I will go into business; I will weave baskets; it is a thing which I can do. [...] He had not discovered that it was necessary for him to make it worth the other's while to buy them, or at least make him think that it was so [...]. (1983, 61)

Thoreau diagnoses different understandings of economic transaction as the problem between the whites and the Natives. In his analysis of the Native's failure to sell his baskets, Thoreau employs capitalist logic by suggesting that appealing to the customer's goodwill is the wrong strategy: the customer needs to believe that the *product* and not the *producer* is worth the money spent. The more desirable the product appears to be, the more money the producer is guaranteed to receive in exchange, regardless of whether or not he is on the verge of starving. Thoreau continues by recounting his own efforts in basket weaving and notes that he undertook this task for pure pleasure, not in order to sell his baskets. He believes that strict business habits are indispensible to every man, however, and he does not regard himself to be an exception. What is important to him is the ability to transact business with the fewest obstacles, by making use of the resources one's environment has to offer, which guarantees his independence from the goodwill and the money of his fellow citizens. In his account, the notion of business thus exceeds a strictly economic meaning and is redefined throughout the text.

Thoreau continues his musings on business and trading with a lengthy discussion of the necessaries of life, namely food, shelter, clothing, and fuel, which he had interrupted earlier to talk about his past life in Concord and the encounter with the Native American basket weaver. It is almost as if he forcefully returns to his reflections on the necessaries of life because he does not want to address the Native American's fear that the incompatibility of doing business with the white man might end in his starvation and death (and in that of his people). By returning to the purpose of his own business transactions and the necessaries of life, Thoreau pushes the Native American towards the fringes of his narrative; very clearly, discussing that man's fate is not important enough and too uncomfortable for Thoreau to linger. Almost halfway through *Walden*, however, in the chapter "The Bean-Field," the presence of the dead, possibly starved Indians comes back to haunt him. When he hoes his bean-field, it suddenly occurs to him that he "disturbed the ashes of unchronicled nations who in primeval years lived under these heavens" and that their "small implements of war and hunting were brought to the light of this modern day" (1983, 204). Thoreau cultivates his beans on the ashes of the Indians' lost culture, but the ruins of their culture are so far removed from him that there is nothing for his imagination to work with—he cannot reconstruct the long lost past underneath his feet, so he immediately abandons the thought and returns to contemplating his present situation.

Later in the book, the Irish similarly resurface unexpectedly but, because their presence is uncomfortable, they are immediately pushed to the back of Thoreau's mind again. When he takes a walk on the pond's shore in winter, Thoreau recalls the raft he had found in the course of the summer, which the Irish had pinned together when the railroad was built and which had been swept ashore with other driftwood. But rather than contemplate the fate of the Irish workers, Thoreau prefers to collect the logs and drag them across the frozen pond, an activity he finds amusing and exhilarating. As it is cold, he burns the logs, which "made a very hot fire" as he contentedly states, thus destroying the last visible sign of the Irish presence at the pond (1983, 297). The Irish meet the same fate as the Native Americans, as they are removed from Thoreau's garden by being turned into ashes. The loss of their presence is not mourned or bemoaned, as Thoreau cannot integrate the objects that testify to the Irish and the Native American presence at the pond into his microcosm in a meaningful way. Similar to the ethnic park workers in *Jurassic Park*, the Irish and the Indians are tolerable casualties in the project of colonization and progress; they are the ghosts that haunt the machine which has forced its way into the garden.

Conclusion

Pastorals such as *Walden* are, as H. Daniel Peck notes, "always defined by what they exclude and by the tensions or 'interruptions' they exhibit in the act of excluding" (1990, 131–132). The idealization of a pure and innocent nature depends upon the disavowal of everything that might interrupt the ideal, which indeed makes repression a defining generic convention of the pastoral. The list of tensions and interruptions in *Walden* is indeed long: Thoreau's peacefulness at Walden is threatened by visions of the future (the railroad), on the one hand, and the specters of the past (Native Americans), on the other. "So is your pastoral life whirled past and away," Thoreau concedes as the presence of the machine forces itself onto the site and succumbs to the machine: "But the bell rings and I must go off the track and let the cars go by" (1983, 167). As he ponders over the changes the railroad brings about, it is clear to Thoreau that the machine will eventually destroy the garden, even though he indicates that Walden Pond had already been destroyed in his lifetime and that, therefore, the ideal he longs for does not exist anymore anyway. The pond of his boyhood, Thoreau recalls, used to be "completely surrounded by thick and lofty pine and oak woods" but is now circled by cleared shores and is employed as a quarry for ice and lumber (1983, 239). In other words, Thoreau's Walden Pond "was no wilderness landscape but instead a colonized space from which Indians and most wildlife had already been removed," as Louise Westling has argued (1993, 148). *Walden* thus constitutes an attempt to negotiate between the landscapes of colonial exploitation and an imagined Garden of Eden, between a state of corruption and a state of innocence, between the realities of imperialism and the myths of a glorious past. As Leo Marx suggests, Thoreau relocates his pastoral hope in its "traditional location," in literature, in his own craft, where his hope can be manifestly realized (1964, 265).

Marx's focus on the pastoral ideal in texts like *Walden* is supposed to shed light on "the American view of life" (1964, 3) and on "American experience" (1964, 4), as distilling distinctively American patterns, symbols, myths, and experiences was essential for the interpretation of America's role in a post-World War II world. Forged during the Cold War, this particular school of practicing American Studies promoted the notion of American Exceptionalism, which could serve to explain and justify America's dominance and its role as the leader of the 'free world.' Marx emphasizes several times that he reads *Walden* as a distinctly *American* version of the romantic pastoral, which implies that, consequently, Thoreau's text can be useful in defining and describing 'the American experience.' Written at a time when the field of American Studies was still attempting to consolidate itself from an uncertain position at the academic periphery, *The Machine in the Garden* became one of the founding studies of the discipline and placed the political at the very center of American Studies. "The machine's sudden entrance into the garden presents a problem that ultimately belongs not to art but to politics," Marx concludes his study, which one can either read as advice to fellow literary scholars not to cross that line into politics or as an indication that art and politics are fundamentally interlocked and that, therefore, the study of all arts is always inevitably also a political project.

Paul Lauter's approach to American Studies certainly follows the latter interpretation, as he argues that doing American Studies "is a kind of political act" (2001, 2). Lauter's analysis of *Jurassic Park* reflects the state of American Studies at the turn of the millennium; he states that American Studies continues to be a contested field of study that permanently redefines itself and adjusts its scope and methods. Considering that his book was published just a month before the events of September 11, which led to yet another shift in how the discipline imagined and defined itself, Lauter's words sound almost prophetic. His reading of *Jurassic Park* is embedded in a brief recollection of how he and other practitioners of American Studies, "have in the last quarter-century deviated from the well-trimmed paths of [Jane] Austen and [Henry] James" and found themselves "hacking through the underbrush of 'Jurassic Park'" (2001, 102). There are several possible readings of Spielberg's film, he notes, which "embody significant shifts in how one does American Studies" but which ultimately all return to "the question of the social functions performed by established academic institutions [...] and their definitions of knowledge" (2001, 102–103).

To Lauter, American Studies functions as a tool for comprehending cultural production, distribution, and consumption, which is why blockbusters such as *Jurassic Park* and the cultural work they accomplish offer themselves as objects of analysis. *Jurassic Park* is a successful product and commodity as well as a catalyst for the selling of other products, and it functions "as a major way in which capitalism in the late twentieth century is rationalized and thus sustained" by making visible "contradictions within commodity capitalism, and then emotionally grounding them in a microcosm of spectacle and pleasure," Lauter explains (2001, 113). A serious analysis of the film thus sheds light on the triumph of commodification and on the 'paradise' of consumption into which the Western world has turned. Furthermore, the worldwide success of films like *Jurassic Park* prompts American Studies to look beyond national borders and to understand American (popular) culture as a global currency. This observation backs up Lauter's assessment of American Studies having become more comparative and global in outlook, more conscious of the political functions of cultural texts, and more interested in the dissemination and appropriation of American culture in different parts of the world (see 2001, 17–19).

As this essay has shown, reading Marx's exploration of *Walden* and Lauter's analysis of *Jurassic Park* side by side offers glimpses at how American Studies was done, defined, and imagined in two very distinct moments in American culture. Moreover, the different versions of 'America' created in in *Walden* and in *Jurassic Park*, as well as the ways in which these versions succeed and fail, tell us what is at issue in a society at a given point in time. *Jurassic Park* is more than the film-version of *Walden* blown up to a Hollywood spectacle—it is the evolution of *Walden*, delineating the logical progression of Thoreau's observations on modern American culture. Even though these two texts respond to the respective, very diverging sociocultural contexts in which they were written, they both grapple with the imperfection of America and its willful subscription to capitalist ideology. However, if not kept in check, capitalist ideology may develop a self-propelling dynamic that is dangerous and destructive to everyone mired in the system. As I have pointed out, both *Walden* and *Jurassic Park* delineate how capitalism thrives on colonialism and the exploitation of indigenous cultures. Moreover, in both texts the literal and figurative removal of native people and exploited workers is essential to the design of an idealized and perfected version of 'America.' But in the end, the ideal cannot hold and the colonization and exploitation of nature, workforce, and native peoples for the sake of maximizing profit will eventually come back to haunt the colonizers. Machine and Garden destroy each other. Ghosts in the Machine prevail.

Works Cited

Balides, Constance (2000). Jurassic Post-Fordism: Tall Tales of Economics in the Theme Park. *Screen*, 42.2, 139–160.

Biskind, Peter (2000). A "World" Apart. In Lester D. Friedman and Brent Notbohm (eds.). *Steven Spielberg: Interviews*, 193–206. Jackson: University of Mississippi Press.

Briggs, Laurie, and Jodi I. Kelber-Kaye (2000). "There is No Unauthorized Breeding in Jurassic Park": Gender and the Uses of Genetics. *NWSA Journal*, 12, 92–113.

Brigham, Ann (2002). Consuming Pleasures of Re/Production: Going Behind the Scenes in Steven Spielberg's Jurassic Park and at Universal Studios Theme Park. *Genders* 36, 53 pars. 09.12.2012 http://www.genders.org/g36/g36_brigham.html.

Brooks, Van Wyck (1918). On Creating a Usable Past. *The Dial*, 64.7, 337–341.

Bukatman, Scott (1993). Dinosaur Bytes. *Art Forum International*, 32.2, 14–15.

Cafaro, Philip (2002). Thoreau's Environmental Ethics in Walden. *Concord Saunter*, 10, 17–63.

Cavell, Stanley (1992). *The Senses of Walden*. Chicago: Chicago University Press.

DeTora, Lisa (2008). "Life Finds A Way": Monstrous Maternities and the Quantum Gaze in Jurassic Park and The Thirteenth Warrior. In Marcelline Block (ed.). *Situating the Feminist Gaze and Spectatorship in Postwar Cinema*, 2–26. Newcastle: Cambridge Scholars Publishing.

Garber, Frederick (1991). *Thoreau's Fable of Inscribing*. Princeton: Princeton University Press.

Gordon, Andrew (2008). *Empire of Dreams: The Science Fiction and Fantasy Films of Steven Spielberg*. Lanham: Rowman & Littlefield Publishers.

Hamscha, Susanne (2013). *The Fiction of America: Performance and the Cultural Imaginary in Literature and Film*. Frankfurt am Main: Campus.

Hawkins, Harriet (1995). *Strange Attractors: Literature, Culture, and Chaos Theory*. New York: Prentice Hall.

Jurassic Park (1993). Dir. Steven Spielberg. Writ. David Koepp and Michael Crichton, based on the novel by Michael Crichton. Perf. Sam Neill, Laura Dern, Jeff

Goldblum, Richard Attenborough, and Samuel L. Jackson. USA: Universal Pictures.

Lauter, Paul (2001). *From Walden Pond to Jurassic Park: Activism, Culture, and American Studies*. Durham: Duke University Press.

Marx, Leo (1964). *The Machine in the Garden: Technology and the Pastoral Ideal in America*. New York: Oxford University Press.

Michaels, Walter Benn (1977). Walden's False Bottoms. *Glyph*, 1, 132–149.

Paryz, Marek (2012). *The Postcolonial and Transcendental Experience in American Transcendentalism*. New York: Palgrave.

Peck, H. Daniel (1990). *Thoreau's Morning Work: Memory and Perception in A Week on the Concord and Merrimack Rivers, the Journal, and Walden*. New Haven: Yale University Press.

Pratt, Mary Louise (2008). *Imperial Eyes: Travel Writing and Transculturation*. New York: Routledge.

Sanford, Charles L. (1965). Review of *The Machine in the Garden*. *American Quarterly*, 17, 272–274.

Shay, Don, and Jody Duncan (1993). *The Making of Jurassic Park*. New York: Ballantine Books.

Spence, James H. (2008). What Is Wrong with Cloning a Dinosaur? Jurassic Park and Nature as a Source of Moral Authority. In Dean A Kowalski (ed.). *Steven Spielberg and Philosophy*, 97–111. Lexington: University Press of Kentucky.

Stern, Megan (2004). Jurassic Park and the Moveable Feast of Science. *Science as Culture*, 13.3, 347–372.

Taylor, Bob Pepperman (1996). *America's Bachelor Uncle: Thoreau and the American Polity*. Lawrence: University Press of Kansas.

Taylor, Philip M. (1992). *Steven Spielberg: The Man, the Movies, and their Meaning*. New York: Continuum.

Thoreau, Henry David (1958). *The Correspondence of Henry David Thoreau*. Eds. Walter Harding and Carl Bode. New York: New York University Press.

— (1983). *Walden and Civil Disobedience*. Ed. Michael Meyer. New York: Penguin Books.

Westling, Louise (1993). Thoreau's Ambivalence Toward Mother Nature. *ISLE: Interdisciplinary Studies in Literature and Environment*, 1.1, 145–150.

Ziff, Larzer (1981). *Literary Democracy: The Declaration of Cultural Independence in America*. New York: The Viking Press.

Rural Electrification plus Government Power: Leo Marx, the Materialist Dialectic, and *Power and the Land*

Marlon Lieber

Leo Marx and the Materialist Dialectic

When Leo Marx reassessed his classic 1964 book *The Machine in the Garden* in 1986, he attested pastoralism "a dialectical mode of perception" (44). This was because it embodied "the need, in virtually all aspects of experience, to mediate—to strive for acceptable if transitory resolutions—between the constraints of society and the constraints of nature" (ibid.). Evoking the dialectic and the concept of mediation at this point can only be meaningful if we understand the relation between those two poles, nature and society, as dynamic. The dialectic as a mutual mediation of "antithetical pairs" (Adorno 1990, 136) means that determinations of each pole are simultaneously present within its opposite. In other words, nature and society are not mutually exclusive but interrelated. What is more, this implies that a change in one of the two has consequences in the other one; they are interdependent. Finally, this leads us to the conclusion that Leo Marx's model—the pastoral design—is able to adapt to changing historical configurations of nature and society. Indeed, he emphasizes himself that studying the "history" of pastoralism "makes sense only if we take account of the distinctive meaning it acquires in each specific social context in which it is reactivated" (1986, 44). Thus, there are neither nature nor society as such but only historically specific forms of their interrelation.

On the other side of the Atlantic, only two years prior to the publication of *The Machine in the Garden*, the German philosopher Alfred Schmidt published his pathbreaking study of *The Concept of Nature in Marx* (based on a dissertation he had written under Max Horkheimer and Theodor W Adorno). On the one hand this study formulated a critique of Soviet orthodoxy and Friedrich Engels' conception of a 'Dialectics of Nature'; on the other it prepared the ground for a more dynamic systematization of the role of nature in Karl Marx's works (especially in his later writings such as the *Grundrisse* and *Capital*). Schmidt's dialectical materialism holds that there is no matter as such "as a supreme principle of being" (2014, 34). Rather, matter and, thus, nature are always subjectively mediated through human praxis. It is, furthermore, only through "enter[ing] into new productive relationships with each other and with physical nature" (ibid., 109) that human beings attain knowledge of nature and its laws. Thus, the relation between subject and object is not a static one but changes in this dialectical process so that "different theoretical reflections" about nature "correspond to the different historical forms of man's struggle with nature" (ibid.). It follows that the role nature plays in the human imagination is constantly changing according to the level of science and technology: "mythology" refers to forms of consciousness that express "the compulsion of uncontrolled physical nature" (ibid., 141), and this is in the historical process replaced by a more enlightened, scientific frame of mind.[1]

Leo Marx seemed to have had a clear, intuitive grasp of this dialectic. In his excursus on Virgil's pastoral poetry that is meant to exemplify the universality of the pastoral design, Marx is clear that the move away from civilization that characterizes the pastoral idyll "stops far short of unimproved, raw nature," where the "wild uncertainties of nature" reign (1964, 22). The pastoral imagination, thus, is already beyond "mythology." Yet, the existence of "raw nature" is a necessary prerequisite for the pastoral dynamic—which otherwise would lose one of its constitutive poles. This is why pastoralism was so attractive for the European settlers who colonized North America: after all, there still seemed to be "unimproved" land in abundance. But just as the harmonious "middle ground" (ibid., 23) is to be distinguished from wilderness, it also cannot be one in which technology and advanced civilization reign supreme. It needs to be noted, however, that Leo Marx acknowledges that it is not technology as such that poses a threat to the pastoral idyll, but that it only turns into a disruptive force in specific social arrangements.[2] To wit, Schmidt pointed out that the

1 We can see the influence of Schmidt's teachers, who claimed that the difference between 'mythical' and 'rational' thinking is one of degree, not of kind; or, as they famously put it, "myth is already enlightenment; and enlightenment reverts to mythology" (Adorno/Horkheimer 1997, xvi).

2 Later, Marx, by way of an examination of Heidegger's conception of technology, laid the philosophical foundation for this argument, distinguishing between "technology itself" and "technocratic modes of behavior" (1984, 652). This distinction between technology as such and the use of technology in particular social settings will return in a later section

"productive relationships" individuals establish are not just ones "with physical nature," but at the same time "with each other". In the first of Virgil's *Eclogues* the shepherd Melibœus has been evicted from his land in Rome, the city that embodies "organized power, authority, restraint, suffering, and disorder" (ibid., 21)—all referring to a particular social formation, not to technology as such. Ultimately, the pastoral synthesis of "the sophisticated order of art and the simple spontaneity of nature" (ibid., 22) means that there should neither be the chaos of "unimproved, raw nature" nor a stunting of the subject's freedom by an oppressive social order.

In line with a dialectical vision, social formations—and particularly the one that Leo Marx is primarily interested in: capitalism—are precisely oppressive and threatening to subjective freedom, because they possess determinations of nature themselves. In other words, the capitalist mode of production amounts to a system that, despite all advances in the sciences, is not under the rational and conscious control of human beings. This is why, in the preface to *Capital*, Karl Marx likened "the development of the economic formation of society" to "a process of natural history" (1990, 92). Michael Heinrich aptly summarizes the object of Marx's critique of political economy: "Within political economy, social relationships such as exchange and commodity production are 'naturalized' and 'reified,' that is, social relationships are conceived of as quasi-natural conditions" so that "it appears as if things have the properties and autonomy of subjects" (2012, 34, emphasis in the original). Marx discloses this "fantastic form" as being based on "the commodity-form, and the value-relation of the products of labour within which it appears" that is, ultimately, "but the definitive social relation between men themselves" (1990, 165). Relevant for us is that even the heretofore most advanced form of human civilization, capitalism, is one in which human beings are not in control of the system that they have created. Interestingly, this motif also appears in Leo Marx's analysis of Emerson, who famously wrote:[3]

Things are in the saddle,

And ride mankind. (quoted in L. Marx, 178)

The dynamic nature, then, of Leo Marx's dialectic of nature and society—that is often reminiscent of the materialist dialectic—assures that his

of this article in Herbert Marcuse's distinction between "technics" and "technology" (1941, 414).

3 On Emerson and reification see Porter (1986, 199–203); see also Voelz (2010, 43–49).

model will continue to keep its relevance—for at least as long as there is no reconciliation between nature and society that is more than "transitory," as Leo Marx had put it. While there might hardly exist any "unimproved, raw nature," the "natural laws of capitalist production" (K. Marx 1990, 91) are still very much with us today. In yet another dialectical turn of events this mode of production, by being "indifferent toward the destruction of the natural foundations of life" and by treating nature "as a mere instrument" for the production of surplus value, will in the future very likely lead to "global economic degradation" (Heinrich 2012, 117). In other words, the "destructive potential" (ibid.) of capitalism will turn nature itself into a destructive force again. The ideal of a "middle ground" in which the destructive potential of both poles, nature and society, is somehow held in check, then, might remain attractive—possibly also for political movements.[4] However, a point that Leo Marx might have emphasized more strongly (although doing so does not contradict his argument) is that the invention of new technologies usually means that older ones lose their force to cause psychic shocks, such as the ones he records in the documents of American writers from the nineteenth century that so often feature the railroad. In fact, the railroad nowadays hardly has a comparable impact; we can rather imagine a somewhat nostalgic feeling being evoked by one, especially if it is a steam locomotive.[5] Again this shows how it is not technology as such that threatens to disrupt the pastoral idyll. In the film to be discussed in the next sections it is precisely a 'machine,' which is needed to save the 'garden.' Thus, the reading of the film will, on the one hand, allow us to see how Leo Marx's analysis of pastoralism remains a useful tool for cultural analysis. On the other hand, keeping in mind Karl Marx's critical theory of the relation between nature and society and their

4 Originally, Marx had ended *The Machine in the Garden* by supposing that pastoralism had lost its power as a basis for political action in the mid-twentieth century (1964, 365). When he reflected upon the fate of his book later, however, he acknowledged a pastoral strain in the American New Left (1986, 60–66). Elsewhere he writes that "[s]ince the Vietnam era, this collective mentality [i.e. pastoralism] has informed many of the dissident movements adhered to by discontented Americans" (1999, 494).

5 That is to say, there is a tendency to naturalize older technologies that human beings have become familiar with. Interestingly, Theodor W. Adorno, in whose aesthetic theory the concept of "technique" plays a central role, makes a similar case for musical techniques. He criticizes those who reject "new music" (i.e. the Schönberg School) because of its alleged "intellectualism" (read: artificiality) for acting "as if the tonal idiom of the past 350 years were itself given by nature and as if it were an attack on nature to go beyond what has been habitually ground in" (2006, 13).

mutual interaction can help us in our attempt to understand why particular cultural artifacts might find specific solutions to imagine this relationship—or why the attempted solutions might fail—and how a given socio-historical context might inform these.

New Deal Electrification and *Power and the Land*

The documentary film *Power and the Land* (1940), directed by Dutch filmmaker Joris Ivens, could arguably be called 'New Deal propaganda.' A concentrated effort to electrify rural America was undertaken in the 1930s, because in the preceding decade 90 percent of U.S. farmers did not have access to electricity—as opposed to Western European countries, where a majority of farms was already electrified (Nye 1990, 287). One problem was that the private utilities usually refused to supply electricity to rural regions and to build distribution lines, because they saw "little profit" (ibid., 292) in serving a farm population that was much more dispersed than their urban customers and that, additionally, not even used much electricity on their farms. Still, when Franklin D. Roosevelt was elected President, he made an effort to advance rural electrification, stating that "[e]lectricity is no longer a luxury, it is a definite necessity" (quoted in ibid., 304). Two newly founded government agencies that were part of the federal reaction to the Great Depression, the Tennessee Valley Authority (TVA) and the Rural Electrification Administration (REA) created in 1933 and 1935, respectively, set out to transform rural America. The idea was for farmers to set up cooperatives that should "build their transmission lines and operate an electrical service themselves" (ibid., 317). The REA provided government loans at a low rate to the cooperatives, and, in exchange, supervised the use that was made of this money (Kline 1997, 21).[6]

6 This article does not purport to give a detailed analysis of the film itself, nor an exhaustive account of its history. Instead, its use of the trope of the machine and the garden and, more precisely, its adaptation of the Jeffersonian ideal discussed in the next section to the conditions of the 1930s, is analyzed by drawing on select scenes from the film. For a historical account of New Deal electrification and a brief discussion of *Power and the Land* see Nye (1990, 287–338). Kline (1997) analyzes the film and the history of its production and reception. William Alexander's history of leftist films in the 1930s discusses Joris Ivens' career and has a section specifically on *Power and the Land* (1981, 279–

Power and the Land, work on which began in 1938, should "induce often-reluctant farm men and women to participate in its government loan programme," and it was also supposed to "garner support from equally skeptical electrical appliance dealers and politicians who thought farmers could not afford electricity" (ibid., 19; see also Alexander 1981, 279–280). The REA at first tried to sign Pare Lorentz as director. Lorentz had already directed two government-sponsored documentaries in the 1930s: *The Plow That Broke the Plains* (1936) and *The River* (1937). The latter, a film about floods that had ruined the land next to the Mississippi river—and about efforts of the TVA to help the people living there—, was awarded the first prize for documentary films at the Venice Film Festival. Lorentz was subsequently made head of the U.S. Film Service (Kline 1997, 20). But since he was busy shooting another film, he was not available as director for *Power and the Land.* Instead, Joris Ivens was signed, who was a Dutch filmmaker whose early films had received critical approval in the U.S., and who had influenced American leftist filmmakers on earlier visits (Alexander 1981, 121). He was also a self-proclaimed communist and had already shot films not only in the Netherlands, but also in the Soviet Union, in China, and in Civil War Spain. This must have been a reason why he seemed to have liked the idea of making a film about the efforts of the government to support the 'common people' against profit-hungry private enterprise. Indeed, "Ivens was for the New Deal" (ibid., 281) and actually wanted to emphasize "the drama of the conflict between farmers and the private utilities" (Ivens quoted in ibid.). However, the outline of the film had already been sketched by Lorentz. Since his U.S. Film Service had been under attack in Congress, he wanted to avoid overtly criticizing private enterprise (ibid., 282), and Ivens had to comply.

Ivens, along with writer Ed Locke and cameraman Arthur Ornitz, found an Ohio family, the Parkinsons, on whom the film was supposed to focus. The idea was to do a before-and-after story centered on their life before electricity and the many advantages that came with electrification.[7] This meant that the film was scripted and mainly reenacted by non-actors

287). Finally, Stott (1973) and Davidson (1983) give accounts of documentary film in the 1930s.

7 Kline points out that many Ohio families actually already possessed "conveniences" such as internal combustion engines, running water, or battery powered radios (1997, 22). Thus, somewhat ironically, he assumes that the film crew actually removed such technologies as the Parkinsons already owned in order "to emphasize the technological extremes in the before-and-after format" (ibid., 24).

(ibid., 280). Also, there is no dialogue, but the film relies on a narration written by the writer Stephen Vincent Benét (who had won a Pulitzer Prize for his narrative poem *John Brown's Body* in 1929). Somewhat problematically, Benét's narration actually leads to a silencing of the farmers themselves. In fact, they do not have a voice of their own but are spoken for, and the narrator's comments also guide the audience's reception of the film. Still, Ivens developed harmonious relations with the family, which he portrayed in a very positive light, in keeping with his lifelong goal "to make his art a part of the dynamic efforts of exploited working people to take pride in their work and to better their conditions" (ibid., 115). The film starts with images of Bill Parkinson, the father of the family, and his two sons who have to get up before the break of dawn in order to feed and milk the cows in the barn that is only illuminated by kerosene lamps. His wife Hazel also has to get up to feed the family and to do various household chores.[8] There is a somewhat dramatic climax when the Parkinsons have to take back and pour away two cans of milk that they had been trying to sell, because they turned sour due to insufficient cooling. After a long day of hard work a visibly tired family meets at the dinner table, again only under the dim light of a kerosene lamp.

More than half of the film (which has a running time of 38 minutes) has by now passed. On another day, after a sequence of collective work, Bill Parkinson and his (male) neighbors discuss the advantages of electricity. Or, more precisely, it is the narrator who speaks: "Now they talk it over; the country way, the slow, cautious decision of the people." We hear the narrator's voice telling us that the farmers want to get electricity, but that the power companies will not supply them. But then one farmer (we do not know whether it is Bill Parkinson) is said to point out, "but I hear there's a new kind of power—government power." In a town-hall meeting the farmers decide to form a cooperative, and soon we see them erecting a pole for the power lines—called a "liberty tree" by the narrator, who compares it to poles planted "back in revolutionary days".[9] The rest of the film

8 The film relies on a very strict sexual division of labor: the women take care of everything within the house (food, laundry, etc.), whereas the men are active outdoors. Interestingly, the efforts of the Roosevelt administration to propagate rural electrification were very often justified by pointing out how much running water, a washing machine, or an electric stove would ease the life of women on the farm (Nye 1990, 303–344).

9 Davidson notes that the "tendency to employ simplistic and patriotic rhetoric persisted throughout the decade; even in the late thirties [...], the multi-dimensionality of American documentary films was diminished because of their insistent, almost obsessive

"depicts the utopia of an electrified farm" (ibid.). While the Parkinsons, especially Bill, are still somewhat fascinated by the new possibilities of all their new appliances—that "appear," as Davidson notes, "somewhat miraculously, in the Parkinson home after the family joins the cooperative" (1983, 82)—they quickly seem to have accepted this radical transformation of their lives and we see them pursue their tasks that have become much less exhausting. "The day ends with a grateful family enjoying a sumptuous meal cooked on an electric range" (Kline 1997, 24). The last shot is a close up of Bill Parkinson with a smile on his face.

Thomas Jefferson's Pastoral Vision and Rural Electrification

Leo Marx calls Thomas Jefferson's vision of agrarian America "no doubt the most nearly official declaration of pastoral in modern history" (1986, 59). What makes Marx's discussion of select passages from Jefferson's *Notes on the State of Virginia* (1785) especially valuable is his acknowledgment of the various layers in Jefferson's argument. That is, instead of simply seeing the celebration of rural life and the independent yeoman farmer as a backward-looking idealization of a life made obsolete by technological progress, Marx records a shift in Jefferson's rhetoric: in Query XIX of the *Notes* (1984, 290–291), there is a change in Jefferson's tone from that of a "scientific rationalist" whose judgments are based on the received teachings of political economy (L. Marx 1964, 124) to "a highly figurative, mythopoeic language" (ibid., 126) that draws on the style of pastoralism. That is to say, Jefferson is "adopting a point of view for which an accepted literary convention is available" (ibid.). Here are Jefferson's famous words:

> Those who labour in the earth are the chosen people of God, if he ever had a chosen people, whose breasts he has made his peculiar deposit for substantial and

reiteration of traditional democratic ideals, often invoked to 'celebrate' the land and its people" (1983, 71). Given elements in *Power and the Land* like the town-hall meeting, the "liberty pole" and its celebration of the pioneer spirit, or the role of the land itself (more on which in the next section), it is safe to assume that this still holds true for this 1940 film. Of course, this might have been a reason why it was *not* considered to be propagandistic even by critics of the New Deal (Kline 1997, 27), and the fact that the narrator's scripted words guide the viewer's experience of the film might have facilitated this critical reaction.

genuine virtue. [...]. Corruption of morals in the mass of cultivators is a phænomenon of which no age nor nation has furnished an example. (1984, 290)

However, later in his life Jefferson himself qualified his pastoral vision. In an 1816 letter to Benjamin Austin he claimed that the "manufacturer" had to be placed "by the side of the agriculturist" (ibid., 1371). Earlier he had already acknowledged that his pastoral vision could exist in "theory only" (quoted in L. Marx 1964, 133). John Kasson similarly points out that Jefferson knew that "America [...] was not immune to the laws of natural and political economy which governed the rest of the world" (1976, 21). Kasson sees Jefferson's subsequent acceptance of manufactures as being based on the conviction that "[r]emaining within an essentially agricultural context instead of clustered in cities such manufactures would not threaten, but rather enhance, the independence and virtue of American society" (ibid., 25). Finally, what this means is that Jefferson himself understood the dialectic of nature and society. The third President of the United States, then, understood that the pastoral idyll was not a fixed rural landscape, but that the idea of a harmonious reconciliation of the forces of nature and society was in need of a constant redefinition based on "ever-changing circumstances" (L. Marx 1964, 139).[10]

Power and the Land clearly draws on Jefferson's understanding of the land as providing Americans with virtue and strength. But the film, too, does not assume that land and human beings—or nature and society—are the poles of a static opposition. In fact, while *Power and the Land* explicitly evokes the image of the healthy energy of the land, it makes clear that an adaptation to changing conditions has to be made. The narrator of the film claims

It's good land, here. Not prime, but good. Long settled land. Bill Parkinson was born here in 1889, the year his father bought the farm. [...]. He used to do general farming, but the land got tired. [...]. Not the best land, not the worst. But it has raised five healthy Americans and Bill and Hazel. That's something for land to do.

Importantly, the land has not just raised anyone, but "healthy" people—which can just as much refer to physical as to moral health. But now "the land got tired," meaning that it is in danger of losing this ability. And, thus,

10 "The controlling principle of Jefferson's politics is not to be found in any fixed image of society. Rather it is dialectical. It lies in his recognition of the constant need to redefine the 'middle landscape' ideal, pushing it ahead, so to speak, into an unknown future to adjust it to ever-changing circumstances" (L. Marx 1964, 139).

the film establishes the idea that a machine, that is, electricity is needed in order to rejuvenate the garden, as it were.[11]

The necessary technology, however, is not technology as such, but technology as employed by independent family farmers—and the film leaves no doubt about this. After the town-hall meeting, where a majority of the farmers has been in favor of forming a cooperative ("There are no private investors, no profit-making," exclaims the narrator, who explains how the REA plans to support the farmers to the audience—i.e., the farmers at the meeting and the farmers who would potentially see this film in a movie theater and who were the primary audience that the film was addressed to), a brief newsreel-like sequence lists the successes of the REA from 1935 to 1940 ("and new lines going up almost everywhere at the rate of 500 miles a day," ends the narration). But suddenly the music assumes a rather threatening, somber tone—and what should we see but a factory with dark clouds of smoke rising from its chimney and covering a transmission tower? "The city, the place that always had power," are the narrator's words. There is a cut to another tower and the camera follows a power line that extends from it. However, when the movement of the camera stops, it settles on a power pole that is placed in a rural landscape. Then, the music becomes rather upbeat and lighthearted again, and we see the farmers raising their own pole. The film, thus, establishes a striking contrast between the use of electricity and other industrial technologies in the city and in the country. While urban industrialization is portrayed as threatening and potentially oppressive and urban power is criticized, rural electrification assumes a liberating air—after all, the pole erected by the farmers is called a "liberty pole" by the narrator. Like Jefferson, who believed in the potential of manufactures as long as they are not "clustered in cities" (Kasson 1976, 25), the film emphatically supports the idea that modern technologies can be used for the benefit of people as long as they do not interfere with their independence.

At the same time, *Power and the Land* makes an effort to distinguish its pastoral vision from "raw nature." However, as we have seen, the threat is not coming from "unimproved" nature, but from a land that has become "tired" and in need of electrification. Still, it fulfills a structurally similar

11 As such it resembles what David Nye calls a "recovery narrative," which is about "remaking a despoiled landscape. [...] To restore natural beauty and environmental harmony, a countervailing social force enters the area in the form of a nonprofit organization or government agency" (2003, 294).

function as the wilderness in Marx's reading of Virgil. Here Melibœus's fate represented the "state of chaos" that exists in wild nature. The evicted shepherd recounts that he had to abandon the newborn kids of one of his goats, which is interpreted by Marx as signifying that the "principle of natural fecundity is threatened" (1964, 21)—because the reproduction of life itself is no longer possible. While the Parkinsons's land is only "tired," the threat a non-electrified farm poses for the young is still emphasized. To wit, we have seen how the family had to pour away two cans of milk, because they were spoiled due to lack of cooling. But even earlier the film had already linked the milk that they produce to the fate of Americans. At the beginning a text is superimposed over the images of the film:

> Here is one American farm, one American family—the Parkinsons, there are thousands like them, all over the union. The bread we eat and the milk we drink depend upon Bill and Hazel Parkinson—their kids and their stock and their farm and how they're making out.

Later we see Bill Parkinson milking his cows, followed by a close-up shot of bright, white milk being poured into a can. The narrator comments, "Milk that makes bones and muscles for the children of a nation. But milk has to be properly cooled." The film, thus, very explicitly characterizes milk as being necessary for children to grow. And since the Parkinson family has synecdochically come to stand for all American farmers that feed the nation, this means that the whole American nation—and especially its children—is in danger, if the countryside is not electrified and milk cannot be cooled properly.[12] However, the introduction of electricity is, thereby, naturalized and less of an alien force that signifies industrial civilization. Rather, by being needed to produce milk "for the children of a nation" it assumes the role of a quasi-national mother. To conclude, then, *Power and the Land* quite openly draws on the "syntax of the middle landscape" (ibid., 120) in evoking the image of a peaceful rural utopia that is threatened from both sides: from a nature that threatens the reproduction of human life and from urban industrial civilization. But, as in Jefferson's

12 The film seems to use what scholars of the Copenhagen School of security studies call the logic of the "*existential threat.*" These threats are constructed "when an issue is presented as posing an existential threat to a designated referent object," which "justifies the use of extraordinary measures to handle them," thereby "open[ing] the way for the state to mobilize, or to take special powers, to handle existential threats" (Buzan/Wæver/De Wilde 1998, 21, emphasis in the original).

"dialectical" vision, this is not incompatible with the introduction of new technologies—provided that they can be made to fit into the pastoral idyll.

Pastoralism and Utopia: The "Machine-herd"

At first, this "hymn to rural life in the Jeffersonian tradition," as Nye calls *Power and the Land* (1990, 328) might appear to be rather conservative—and in many ways it certainly is. Its gender politics are hopelessly outdated by our early 21st century Western standards. In addition, the film's suppression of all references to the Great Depression and class struggle make it appear as if the only problem of the farmers was, indeed, the "tired" land—instead of the drastic fall in prices for agrarian goods they had experienced in the preceding decade (Kemp 1990, 17). And we have already seen that Joris Ivens had actually wanted to highlight the conflict between farmers and private utilities but had to refrain from doing so due to pressure from Lorentz. Finally, help comes only through paternalist "government power," as the narrator has one farmer put it in the film. Of course, given the fact that it is a film commissioned by the government itself, most of this should not come as a real surprise. But then, if it is undeniable that the film could have—"in theory," to use Jefferson's phrase—been much more radical, it seems worthwhile to critically analyze the utopian moments in the film as it is—and, ultimately, its limitation—, which is what the remainder of this article sets out to do.

There is the classic Marxist formulation of the idea that technological progress will lead to social progress, or, to be more precise, that the development of "the material productive forces of society" will at one point lead to an overthrow of "the existing relations of production," namely when the latter have become the "fetters" of the former (K. Marx 1994, 211). That is to say, when the forces of production allow for the creation of wealth to such an extent that the private appropriation of it is no longer rational, a "social revolution" (ibid.) should abolish the old order.[13] While it has been

13 This is not to say that Karl Marx regarded this to be a historical law in the strict sense, nor that he considered the dialectic of productive forces and relations of production to be central for his critique of capitalism. Moishe Postone (1993, 9 *et passim*) convincingly argues that this was not the case. However, historically this dialectic has been central in

a vexing question in Marxist scholarship whether this amounts to a historical mechanism that works 'automatically,' it certainly played a role in the young Soviet Union in the 1920s, where industrialization—read: development of the forces of production—was enforced to effect a transition to full communism (often at enormous human costs and, arguably, ultimately without success); this is summarized in Lenin's famous phrase, "Communism is Soviet power plus the electrification of the whole country" (1920). It is safe to assume that Joris Ivens, who by the time of his involvement with *Power and the Land* had already shot *Song of Heroes* (1932) in the Soviet Union, was familiar with this view. According to Kline, it is "[m]ore than likely" that Ivens "had faith in the progressive ideology of electricity" (1997, 23) implicit in Lenin's statement.

But there is a difference between the kind of forced industrialization that we now associate with Soviet electrification and the democratic process that *Power and the Land* depicts. After all, the decision to form a cooperative in the film is portrayed as being made at an old-fashioned (and rather grassroots-democratic) town-hall meeting.[14] There is again a Jeffersonian moment: electricity is supposed to rescue independent farming and make possible the "joy of independence" (L. Marx 1964, 128). Leo Marx's interpretation of Query XIX in the *Notes* holds that "economic criteria" are not what counts as a basis for action for Jefferson (ibid., 126). Instead, it is "happiness" (Jefferson 1984, 291) that is the desired result. And yet there is an example for a utopian vision similar to that of *Power and the Land* which is more contemporary than Thomas Jefferson's writings. The American philosopher Lewis Mumford had published his *Technics and Civilization* in 1934, in which he gives a historical account of technological progress that ends with the "assimilation of the machine." Now, according to Mumford, human beings should finally be able to use the benefits of technology in

---

what Postone calls "traditional Marxism" and has served as a basis for "historical materialism" (Heinrich 2012, 200).

14 We can think of Friedrich Pollock's theory of "state capitalism." The economist, a member of the Institut für Sozialforschung and a close friend of Max Horkheimer, had published an article in the institute's journal in 1941, in which he argued that a new phase in the history of capitalism had replaced the liberal era of the free market. This was to be characterized by increased state control of the economy. If the general plan required it, the government could override the interests of individual capitalists (say, the private utilities that did not want to supply farmers with electricity). However, Pollock distinguished between a totalitarian and a "democratic form of state capitalism." The latter is still "controlled by the people" (1941, 202), and Pollock had the United States in mind.

order to facilitate their lives, without being oppressed or alienated by machines. Jesse Ramírez has recently pointed out how Mumford's writing influenced the German philosopher Herbert Marcuse, who discussed *Technics and Civilization* in a 1941 article for the *Studies in Philosophy and Social Science* (which was the English title of the *Zeitschrift für Sozialforschung*, the theoretical organ of the Frankfurt institute that had moved to the U.S. by then). It is here that Marcuse made the distinction between "technics" and "technology" (1941, 414). Whereas the former refers to the various "technical apparatus[es]" and is, thus, neutral, the latter designates "an instrument for control and domination" (ibid.).[15] In Mumford's writing, however, Marcuse found an idea of technics used not in the service of domination—and one that could even fuel what Ramírez calls a "postcapitalist vision" (2012, 33). The crucial factor that should lead to "radically reduced labor time, the elimination of scarcity, and a true break in the history of civilization and human nature" was to be "American automation" (ibid., 35).

Mumford assumed that the development of technology had by then proceeded so far as to make possible an "economy of surplus" (1963, 401). Thus, the amount of labor time that each individual had to perform could be radically reduced (ibid., 405) because of automation. The goal of this is, as in Jefferson, happiness. That is, Mumford distinguishes between a "capitalist economy [...] under the spell of the profit motive," in which production [is] an end in itself" (ibid., 400), and "production for the sake of life" (ibid., 410), in other words, a way of organizing collective life that is based on the actual needs of the individuals that form it. By not having to spend long hours working, human beings would be free to develop their personalities without restraint. This is also the promise that Marcuse recognizes:

> Technological progress would make it possible to decrease the time and energy spent in the production of the necessities of life, and a gradual reduction of scarcity and abolition of competitive pursuits could permit the self to develop from its natural roots. The less time and energy man has to expand in maintaining his life and that of society, the greater the possibility that he can 'individualize' the sphere of human realization. [...]. No longer chained to competitive efficiency, the self could grow in the realm of satisfaction. (1941, 438)

15 This, in many ways, prepares the ground for Marcuse's extended critique of "technological rationality," which he would develop in *One-Dimensional Man* (1964, xvii).

It is not hard to see that for all three, Marcuse, Mumford, and Jefferson, economic criteria (the "profit motive" or "competitive efficiency") should not determine the way human beings arrange their collective life. At the same time, it is far from any collectivist vision, because the ultimate goal for all of them is the free development of the human personality and "happiness."

Finally, Mumford and Marcuse's vision amounts to a "postcapitalist" utopia because, as Ramírez points out, the idea of automation is able to transcend value (2012, 46). Moishe Postone's re-reading of Marx highlights the centrality of labor in its specifically capitalist form, which is value-producing labor.[16] Crucially, it is only human labor that creates value. Thus, in Ramírez' words

> advanced automation makes value (and thus capitalism itself) impossible. Full automation eliminates the proletarian labor time upon which value is founded. On the view shared by Marx and Marcuse [...], capitalism's true demise occurs not with the socialization of wealth, but with a revolution in the mode of production that abolishes value and the proletarian labor time that fuels it, replacing the latter with a more cooperative form of social production. (ibid.)

What remains to do for the "postcapitalist" worker? Strikingly, Lewis Mumford draws on pastoral imagery in his discussion of work under conditions of automation: the worker does not have to take a "direct part in the process of production," because his or her task will be to watch and to control the machines that do the work. The worker, then,

> is, so to say, a *machine-herd*, attending to the welfare of a flock of machines which do the actual work; at best, he feeds them, oils them, mends them when they break down, while the work itself is as remote from his province as is the digestion which fattens the sheep looked after by the shepherd. (1963, 410–411, my emphasis)

Mumford carries over the ideal of the shepherd—and his qualification "so to say" implies that he knows that he is drawing on a literary predecessor and will, thus, like Jefferson switch to a different, "highly figurative" (L. Marx 1964, 126) language—into the machine age. Even here, the pastoral ideal of a reconciliation of nature and technology exemplified by the figure

16 Thus, Marx can claim in *Capital* that his discovery of the "dual character of the labor embodied in commodities," that is, as use-value *and* value-producing labor, "is crucial to an understanding of capital" (1990, 131–132). This is also why he begins his presentation/critique of the categories of political economy with the commodity and the value form. See also Heinrich's outstanding *Introduction* (2012).

of the "machine-herd" remains attractive—and, if only at times, *Power and the Land* evokes that image. Before the final dinner scene, we can see Bill Parkinson in his barn calibrating and controlling the new machines. Says the narrator, "We've got new tools for the old toil. We learn their ways and set them to work—set them to work to help us all." Labor is no longer alienated and the machine is there to help the people who use it, in order for them to enjoy a fuller life.

The Law of Value and the Failure of Utopia

But our experience of the better part of the twentieth century shows that this utopia of small, independent family farming has hardly become reality—if it ever truly had the potential to be one. We have seen that for Karl Marx value-producing labor was at the very core of the capitalist mode of production. According to Postone, it was one of the failures of "traditional Marxism" not to grasp this, which condemned all attempts to overcome it so far. Rather, such attempts were characterized by the belief that industrial methods of production and value-producing labor could be maintained and that only the social product had to be distributed differently (1993, 9). *Power and the Land*, ultimately, also proves unable to fully imagine a world beyond economic value, despite its evocation of a figure similar to Mumford's "machine-herd." When the Parkinsons have finally received all their new appliances, we see Bill Parkinson inspecting their new water pump in the basement. The narrator says, "Electricity means running water on the farm. You can throw the old pump handle away. This pump works all the time for a few cents a day." Of course, human labor would hardly be that cheap. Thus, the film, which almost envisioned a world in which "pecuniary interests" (Mumford 1963, 356) are no longer relevant and production is no longer undertaken for the production of value but for the happiness of human beings, again reminds us of the fact that from the standpoint of a society based on value-producing labor, machines are more productive and cheaper than living labor. And this means that, like the "old pump handle"—or the kerosene lamp that is replaced by electric light—the Parkinsons might themselves be soon replaced by machines. Karl Marx points out that competition forces individual producers to replace the living labor of workers with machines that are more productive—whether they like it or not. While the Parkinsons might be independent farmers, that is, not employees of a capitalist, their products would still have to compete with those of larger agribusinesses on the marketplace. And because larger corporations were rather able to increase their productivity by investing in new machines, a small family would at one point be very likely to lose in this competition. Then, they would—in a kind of "primitive accumulation" (K. Marx 1990, 873–876)—be forced to sell their labor power as workers; indeed, David Harvey writes that "[i]n the United States, the family farm and small shopkeepers have been major targets for proletarianisation since the 1930s" (2011, 60).

The film, then, unwillingly prepares the ground for a narrative of New Deal electrification that David Nye calls "ironic" (1997, 61). He proceeds from the observation that there has been a significant "shift away from agrarian life" (ibid.) in the U.S., despite the measures taken to save it. He explains:

> Productivity per farm had to increase tremendously for such a change to take place. In this long perspective, rural electrification exemplifies the substitution of capital equipment for labor, bringing both reductions in waste (for example through refrigeration) and improvements in productivity. The irony enfolded within this success is, of course, that rural electrification was not implemented to drive people off the land but to keep them there, by bringing to the country the amenities enjoyed in the city. Yet paradoxically, rural electrification meant fewer farmers were needed. (ibid., 62)

We need only think of great agribusinesses today to see that the utopian moments of *Power and the Land* could not assert itself against the "natural laws of capitalist production" (K. Marx 1990, 91). What this also reminds us of is the limited nature of the intervention that Roosevelt's New Deal constituted. As Tom Kemp point out, Roosevelt's "object was to save capitalism, not to supersede it" (1990, 85).[17] Behind the New Deal programs was the belief that "lack of purchasing power was a causative factor in the depression" (ibid., 86), in other words, that demand was too low. Keeping this in mind, we can see what was also at stake in the program of rural electrification: the farm population, without electricity, previously had had no need for a host of consumer goods such as refrigerators, electric washing machines etc. Electrifying these farms also amounted to integrating this part of the population into the national market as customers, ef-

17 This would also be the time to remember that Thomas Jefferson was not only interested in "happiness" but also in "permanence of government" (1984, 291).

fecting a transition "from independent producer to mass-market consumer" (Nye 1990, 331). But this, finally, had little to do with the goal of "happiness" or "production for the sake of life." Instead, it was about a "revival of profitability" (Kemp 1990, 86). The reason for this is that the utopian goal of production beyond value was not seriously pursued.

Where does that leave us with regard to Leo Marx? While Mumford's unexpected evocation of the pastoral "machine-herd" reminds us that pastoralism is very much compatible with the advanced machine age, insofar as it still can project a resolution "between the constraints of society and the constraints of nature" (L. Marx 1986, 44) in its combination of an "economy of surplus" and a commitment to the idea of "quality of work" (Mumford 1963, 411), we can also see that the dialectic of 'machine' and 'garden' does not find a happy ending as long as the "profit motive" still triumphs over human happiness—which is necessarily the case in a mode of production based on the production of value. Ultimately, then, pastoralism might serve as an anticipation of a better state of affairs. To get there from here, however, we might just need the critique of political economy and forms of praxis based on it.

Works Cited

Adorno, Theodor W. (1990). *Negative Dialectics*. Trans. E.B. Aston. London: Routledge.

Adorno, Theodor W. (2006). *Philosophy of New Music*. Trans. Robert Hullot-Kentor. Minneapolis: University of Minnesota Press.

Adorno, Theodor W., and Max Horkheimer (1997). *Dialectic of Enlightenment*. Trans. John Cumming. London: Verso.

Alexander, William (1981). *Films on the Left: American Documentary Film from 1931 to 1942*. Princeton: Princeton University Press.

Buzan, Barry, Ole Wæver, and Jaap de Wilde (1998). *Security: A New Framework for Analysis*. Boulder: Lynne Riener Publishers.

Davidson, David (1983). Depression America and the Rise of the Social Documentary Film. *Chicago Review*, 34.1, 68–88.

Harvey, David (2011). *The Enigma of Capital and the Crises of Capitalism*. London: Profile Books.

Heinrich, Michael (2012). *An Introduction to the Three Volumes of Marx's Capital*. Trans. Alexander Locascio. New York: Monthly Review Press.

Jefferson, Thomas (1984). *Writings*. New York: Library of America.

Kasson, John F. (1976). *Civilizing the Machine: Technology and Republican Values in America, 1776–1900*. Harmondsworth: Penguin Books.

Kemp, Tom (1990). *The Climax of Capitalism: The U.S. Economy in the Twentieth Century*. London: Longman.

Kline, Ronald R. (1997). Ideology and the New Deal "Fact Film" Power and the Land. *Public Understanding of Science*, 6, 19–30.

Lenin, V.I. (1920). Our Foreign and Domestic Position and Party Tasks. Speech Delivered to the Moscow Gubernia Conference of the R.C.P. (B.), November 21 in Moscow. 29.03.2014 http://www.marxists.org/archive/lenin/ works/1920/nov/21.htm.

Marcuse, Herbert (1941). Some Social Implications of Modern Technology. *Studies in Philosophy and Social Science* [=*Zeitschrift für Sozialforschung*], IX, 414–439.

Marcuse, Herbert (1964). *One-Dimensional Man: Studies in the Ideology of Advanced Industrial Society*. Boston: Beacon Press.

Marx, Karl (1990). *Capital: A Critique of Political Economy*, Vol. One. Trans. Ben Fowkes. London: Penguin Books.

Marx, Karl (1994). Preface. *A Contribution to the Critique of Political Economy*. Trans. Salo Ryazanskaya. In K. Marx. *Selected Writings*. Ed. Lawrence H. Simon, 209–213. Indianapolis: Hackett Publishing Company.

Marx, Leo (1964). *The Machine in the Garden: Technology and the Pastoral Ideal in America*. New York: Oxford University Press.

Marx, Leo (1984). On Heidegger's Conception of "Technology" and Its Historical Validity. *The Massachusetts Review*, 25.4, 638–652.

Marx, Leo (1986). Pastoralism in America. In Sacvan Becovitch and Myra Jehlen (eds.) *Ideology and Classic American Literature*, 36–69. Cambridge: Cambridge University Press.

Marx, Leo (1999). Afterword: The Machine in the Garden. *The Massachusetts Review*, 40.4, 483–496.

Mumford, Lewis (1963). *Technics and Civilization*. San Diego: Harcourt Brace & Company.

Nye, David E. (1990). *Electrifying America: Social Meanings of a New Technology, 1880–1940*. Cambridge: The MIT Press.

Nye, David E. (1997). *Narratives and Spaces: Technology and Narratives of New Beginnings*. Cambridge: The MIT Press.

Nye, David E. (2003). *America as Second Creation: Technology and Narratives of New Beginnings*. Cambridge: The MIT Press.

Pollock, Friedrich (1941). State Capitalism: Its Possibilities and Limitations. *Studies in Philosophy and Social Science* [=*Zeitschrift für Sozialforschung*], IX, 200–225.

Porter, Carolyn (1986). Reification and American Literature. In Sacvan Bercovitch and Myra Jehlen (eds.). *Ideology and Classic American Literature*, 188–217. Cambridge: Cambridge University Press.

Postone, Moishe (1993). *Time, Labor, and Social Domination: A Reinterpretation of Marx's Critical Theory*. Cambridge: Cambridge University Press.

Ramírez, J. Jesse (2012). Marcuse Among the Technocrats: America, Automation, and Postcapitalist Utopias, 1900–1941. *Amerikastudien/American Studies*, 57.1, 31–50.

Schmidt, Alfred (2014). *The Concept of Nature in Marx*. Trans. Ben Fowkes. London: Verso.

Stott, William (1973). *Documentary Expression and Thirties America*. New York: Oxford University Press.

Voelz, Johannes (2010). *Transcendental Resistance: The New Americanists and Emerson's Challenge*. Hanover: University Press of New England.

Putting Machines into Gardens: Walter De Maria's *Lightning Field*, Robert Smithson's *Spiral Jetty* and the Pastoral Imagination*

Laura Bieger

Pastoral Resonances

The Lightning Field: a grid of 400 highly polished steel-poles measuring one mile by one kilometer that has been imposed onto a patch of remote, uncultivated 'wilderness' in the high desert of New Mexico. *Spiral Jetty*: a sculpture made from 6,650 tons of basalt rock and earth that stretches counter-clockwise into the translucent red water of Utah's Great Salt Lake. Completed in 1977 and 1970 respectively, *The Lightning Field* and *Spiral Jetty* are the two best-known works of American land or earth art, a movement that took shape in the late 1960s.[1] They have indeed become so iconic that their photographs cover volumes on twentieth century American art. The artists responsible for them, Walter De Maria and Robert Smithson, gained their fame as trailblazers of postmodern American art. Rejecting the high modernism emblemized by Abstract Expressionism, which had turned New York into the center of the art world in the postwar years, they expanded on the institutional critique forwarded by the avant-garde movements of the 1960s and pushed their programmatic delimitation of the traditional work of art into new dimensions.[2] Yet in doing so, both De

* This essay draws from research that I was able to do on a travel grant provided by the Terra Foundation for American Art.

1 The terms 'land art' and 'earth art' emerged simultaneously. When referring to the art movement as such I will use the term 'land art' because it is the relation of *The Lightning Field* and *Spiral Jetty* to the land they occupy that I am particularly interested in. I will not, however, use the term 'environmental art,' which was circulating in the same semantic field for a while since both works are not concerned with the land they occupy in a sense that would justify the term.

2 The view that the painters who are today associated with Abstract Expressionism epitomize the struggle for a truly modern art was popularized by Clement Greenberg, for example in his landmark essay "American-type Painting." Two important accounts of the postmodern delimitation of art that culminates in the movement of land art are Lippard (1997) and Rosenberg (1972).

Maria and Smithson remained deeply invested with the pastoral imagination that took shape when the spaces to which they turned to realize their signature works were first exposed to 'Western civilization.' To speak with Leo Marx's famous trope, both artworks thrive on putting machines into gardens. But they do so in entirely different ways—that is, they put vastly different machines into vastly different gardens.

I want to begin unpacking their mutual yet vastly different engagement with the pastoral by turning to the emergence of land art within the post-modern redefinition of art that culminated in the 1960s. It had three main trajectories: contenting the idea of the artist as a genuine creator of art objects (sculptures, paintings); contesting the idea of artistic value residing in these objects, especially in their formal properties; and contesting the idea of the gallery or the museum as neutral spaces of display for these objects. Minimal and concept art, the two movements that were most crucial in paving the way for the emergence of land art at the end of the 1960s, dissolved the traditional "objecthood" of art (to evoke Michael Fried's famous term) and the romantic cult of artistic creation attached to it by turning to cheap, non-artistic materials and modes of industrial production, and by shifting the presumed 'essence' of art away from its material realization toward the immaterial spheres of experiences and ideas (the terms concept art and idea art were indeed used synonymously for a while).[3]

They challenged the neutrality of the 'white cube' in which modern art had flourished—and this third trajectory is of particular importance here—by exposing the ways in which this space privileged a cool and disengaged vision ('the disembodied eye') through the highly performative (Fried would say "theatrical") demands of fully embodied reception. If a sculpture sits on the floor rather than on a pedestal and a drawing is applied directly onto a wall rather than onto a canvass it becomes unclear where the artwork ends and the exhibition space begins—leaving it to the recipient to negotiate the boundary between the two by moving around in that space and feeling it out with all the senses (hence the paradigmatic shifts from 'object' to 'subject,' from 'timelessness' to 'duration,' and from a 'formalist' to a 'phenomenological' mode of reception that Fried so lucidly describes in his influential essay).[4]

---

3 See Fried (1995).

4 While Fried (1995) described these features of minimal art (which he dismisses as 'literalist') to expose its flaws and weaknesses, he still produced a highly pertinent analysis of

The most important contribution that land art made to the postmodern redefinition of art can be summarized in one single term: site-specificity. It designates an artistic practice that thoroughly intertwines its work with the sites it occupies.[5] In Walter De Maria's programmatic words, "The land is not the setting for the work but a part of the work" (2010, 118). *The Lightning Field* and *Spiral Jetty* are specific to their sites in at least two intersecting ways: they draw from their 'earth,' that is, from its raw materiality (a dependency that is underscored by replacing the term 'artwork' with 'earthwork'), and they draw from their 'land,' that is, from the presumably desinterested, wide-open spaces surrounding them. And yet the spaces that these works occupy are neither empty nor historically neutral, just as it would be wrong to treat the works themselves as "oversized relocations of sculpture from the gallery to the similarly rarified spaces of the American West" (Roberts 2004, 1–4, here 4). Edward Casey's definition of 'land' as making the 'earth' visible by bringing it into the realm of representation is instructive here. For him, "Land turns the earth inside out, as it were, putting its material contents on display, setting them out in particular places so as to become subject to articulation in language and to play a role in the history of those who live on it" (Casey 2004, 236, quoted in Boetzkes 2010, 15–16).

What the land occupied by these artworks most dramatically puts on display is remoteness. It is indeed this quality that turns this land into such a welcome laboratory for postmodern artistic production. Yet while its remoteness may be a geographical fact that persists even in times of mass air-travel, it is also a cultural fiction that has been crafted over centuries. And if an artwork is located in or even 'made out of' this land, the imaginary, mythical dimension that is instrumental in making it legible and meaningful will be substantial to our encounter with it.

---

this kind of art. With its decidedly somatic appeal to movement and feeling, the minimalist challenge to the 'disembodied eye' was very much a phenomenological one. For further reading on these topics see Sontag 1966, Lippard 1997, Battcock 1973, O'Doherty 1999. For a good, comprehensive overview of art in the 1960s see Sandler 1988.

5 For an early appraisal of site-specific art see Krauss 1979. Looking back at the evolution of this artistic practice in the late 1990s, Kwon describes a development from a strictly place-bound via a social-institutional to a discursive version of it (1997). For a recent account that brings this practice into a dialogue with philosophical (mostly phenomenological) questions about 'earth', 'land' and perception, see Boetzkes 2010.

Most likely, our first encounter with *The Lightning Field* and *Spiral Jetty* will not be in situ but through photographs. The point is hardly peripheral: site-specific art has created receptive conditions that *demand* documentation, both visual and discursive, to make it known beyond its site. To constitute itself as art it needs compelling mediation as much as it needs a specific location—both of which have profound effects on "the way art [is] seen and read as image, as well as the ways in which it [is] talked about" (McFadden 2009, 80).[6] But land art not only unsettles traditional boundaries between artwork and documentation. It challenges our critical assessment of what we see because its dependency on documentation exposes a complex history of the American landscape and its images that "plays off 'resonant mythologies' while constantly creating new ones as well" (ibid.).[7] Yet while it is certainly true that the "American landscape is layered with cultural images [...] dense with myth and meaning" (ibid.), the mythologies resonating through these images are not as blurry as one might suspect. In fact, there is one myth of the land ringing loud and clear from both *The Lightning Field* and *Spiral Jetty* (if not from all major works of land art) that is quintessential to their status as signature works of American art: the pastoral myth that Leo Marx has so forcefully described in his landmark study *The Machine in the Garden.*

For Marx, the pastoralism that took shape in the U.S. under the impact of industrialization is inadequately described with the idea of the 'virgin land,' the concept used by earlier proponents of the Myth and Symbol School, most famously by Marx's teacher and mentor Henry Nash Smith in his influential study of the same title. Detecting the emergence of a "post-romantic, industrial version of the pastoral design" (Marx 1964, 32) in mid-nineteenth century America, Marx distinguishes between a 'senti-

6 For Dave Hickey, the new relation between artworks and their documentation/mediation brought about by the specific conditions of reception is even more crucial than the manipulation of concrete material. "Now there is an art form ideally suited to representation via magazine, work consisting of photographs and documentation that is not presented by journalism but as journalism—a higher form, needless to say. [...] Should these art forms flourish and develop, we shall soon need a kind of National Geographic for Esthetes. [...] Already Philip Leider and Diane Waldman have returned with (literally and figuratively) breathless accounts. New styles of criticism are evolving: it is goodbye Clement Greenberg and Michael Fried, hello Ernie Pyle and Richard Harding Davis [two famous war correspondents]. [...] An artist who makes documents needs an editor, not a dealer" (Hickey 1971, 48). Hickey's essay is still one of the most insightful pieces of writing about the relation of site-specific art and its images.

7 The notion of 'resonant mythologies' is drawn from Hickey 1971, 48.

mental pastoralism' that imagines the machine (Marx's shorthand for the rise of industrialism) as a simple threat to unspoiled nature and a 'complex pastoralism' that imagines the machine as a highly ambiguous force, at times organically welded into the garden as a harbinger of its cultivation, at times threatening to spoil it.[8] Yet while *The Machine in the Garden* locates the temporal center of this development in the nineteenth century and assumes its demise in the era of high technology that Marx saw maturing in mid-twentieth century America, his later writings reverse this judgment, arguing that the pastoral imagination regained momentum in giving voice and form to the dissident critique of the 1960s and in propelling the environmental consciousness emerging at this time (see for example Marx 1970 and 1999/2000).

Yet contemporaneous as they were, the pastoralism that helped articulate the countercultural spirit of renewal does not seamlessly translate into the art projects at stake here. Much of the period's art was, indeed, strikingly removed from the political battles of its time, and land art was no exception.[9] Smithson, for example, rejected the notion of the artist as political activist—"I am not a salvationist"—along with the ecology movement, for him a product of a "moral confusion" that charged concerns about nature with a human desire for post-mortal continuance (Smithson 1970/1996, 239). His earthworks were to unsettle what he perceived as a misguided anthropocentrism at the heart of these nature longings and for which the pastoral imagination was the primary vessel.[10] Yet even if Smithson made a conscious effort to put both his art and writing—he was also a prolific essayist—into the service of de-romanticizing the natural world, I argue that his best-known work is more closely entangled with the pastoral imagination than statements like these would make us suspect.

8 Marx develops the distinction between sentimental and complex pastoralism in the first chapter of his book. Later generations of American studies scholars have severely criticized his work for its narrow focus on canonical—white, male—authors, a reproach to which Marx responds in "Ur-Theory" (2005).

9 Auping discusses the ambiguous ecological politics of land art in "Earth Art" (1983).

10 Smithson's most controversial ideas in this regard remained unrealized as the artist had to give up his plans to cover Miami Islet, a small island in the Strait of Georgia near Vancouver which the local government had already allowed him to use, with broken glass when local chapters of the conservationist movement began to mobilize against the project. For an elaborate discussion of the Island of Broken Glass controversy and Smithson's ambiguous relation to the ecology movement see Graziani 2004, particularly chapter 3.

De Maria's distance from environmental concerns was less outspoken than that of his colleague. Notoriously reclusive and highly selective in his public commentary, he conceived *The Lighting Field* as a laboratory of the senses—"Isolation is the essence of Land Art" (2010, 120)—designed to create profound experiences. Early criticism has often been frustrated with De Maria's rigid control of his work while more recent accounts have shifted to regarding this aspect of his work as part of his position as a conceptual artist.[11] But in any case it is impossible to dismiss the artist's close involvement with aesthetic rather than activist concerns, and again, it is through an implicit pastoralism that this involvement gained force and contour in his work.

The Lightning Field as 'Middle Landscape'

In 1960, almost two decades before completing *The Lightning Field*, Walter De Maria published a short text that celebrates natural disasters as "the highest form of art possible to experience" (1988, 45):

> Put the best object you know next to the grand canyon, the niagara falls, the red woods.
>
> The big things always win.
>
> Now just think of a flood, forest fire, tornado, earthquake, Typhoon, sand storm.
>
> Think of the breaking of Ice jams. Crunch.
>
> If all the people who go to museums could just feel an earthquake.
>
> Not to mention the sky or the ocean.
>
> But it is in the unpredictable disasters that the highest forms are realized.
>
> They are rare and we should be thankful for them. (De Maria 1988, 45)

11 The most outspoken critique of De Maria's authoritarian control over his signature work stems from Beardsley (1981). Both Cooke (2008) and McFadden (2009) have recently set out to revaluate De Maria's work by stressing his admirable consistency as a conceptual artist.

Fig. 1: The Lightning Field *and its promise of profound experience, captured by John Cliett for Walter De Maria.*

From: Szeemann, Harald (ed.). (1999). *Walter de Maria: The 2000 Sculpture*, Exhib. Cat., Zürich: Kunsthaus Zürich, 31.

In what reads both like a piece of conceptual art and an early blueprint for his signature work, De Maria spells out his fascination with profound experiences. His plea to discard the traditional 'objecthood' of art and replace it with disastrous nature is paired with a longing for the real that is to be redeemed through the aesthetic force of the sublime. The most famous photograph of *The Lightning Field* captures precisely this scenario (fig. 1). The steel poles out of which it is made stand out against a large sky that has been dramatically darkened by a thunderstorm as an extremely bright and powerful strike of lightning hits one of them, illuminating it top to bottom. And while the beholder's attention is clearly drawn to this spectacular event it is impossible to miss that the poles placed within this vast and empty space in the rectangular shape of a grid to evoke this 'disaster' have been industrially produced. Those who have read the short text accompanying the publication of this iconic photograph—the authoritative, frequently quoted source of information about the work and only reading provided on site—may also know that the poles "were constructed of type

304 stainless steel tubing with an outside diameter of two inches"; that they have "concrete foundations, set one foot below the surface of the land, [which] are three feet deep and one foot in diameter"; that the grid laid out by their arrangement measures exactly one mile by one kilometer; and that calculating their exact position took five entire months (De Maria 2010, 119).[12] But even without knowing all these things it is impossible to not note that *The Lightning Field* was constructed with heavy machinery and high precision technology—that the 'disaster' that has been so artfully put on display by this iconic photograph is not merely man- but machine-made.

While it is true that thunderstorms frequently occur in this region—"3 per 30 days during the lightning season" (De Maria 2010, 120)—and that the site has been selected precisely for this reason, the scenario depicted in the photograph is thoroughly artificial. Its maker, John Cliett, whom De Maria had hired for the task of visual documentation soon after the construction process was completed, spent weeks on site to capture this motif, using extremely long exposure to create his powerful images (one may indeed think of the photographer's camera as part of the machine placed in the garden here). Other photographs depict their object in more serene states, with less intimidating strikes of lightning and a brighter, more inviting landscape whose rugged vegetation glows in an intense, pasture-like green, or by capturing the magical moment in which the poles are illuminated in shades of orange, pink and purple in the low light of the rising or the setting sun. Yet no matter how different the state or mood of the field, all of Cliett's photographs have a similar way of treating the contrast between the natural and the industrial components of the work. Rather than staging it in ways that would make them appear irreconcilable—the machine brutally intruding on raw, unspoiled nature—they toy with amalgamating the two. The picture that goes furthest in this regard places the field of poles in a distant middle ground where it seems to organically 'grow' out

12 The text was initially published in *Artforum* in April 1980 together with a small selection of John Cliett's photographs. It is the only publication by De Maria about his work and has been controversially discussed for the artist's insistence on separating it with a monochrome sky-blue page from the rest of the magazine. The spread of nine pages devoted to his work was further set apart by including them without the usual pagination. McFadden discusses the publication as one of the artist's several photographic essays, which are, for her, an essential yet so far underrated part of his work that allows us to better understand his consistency as a conceptual artist. See McFadden (2009, 80–87) for the discussion of the photographic essay concerning *The Lightning Field*.

Fig. 2: The Lightning Field *as organic part of the landscape.*

Source: Tiberghien, Gilles A. (1993). *Land Art.* Paris: Édition Carré, 220.

of the soil and belong to the land just as the patches of grass in the foreground (fig. 2).

It is worth pondering over these photographs for another moment, not least because any visitor to *The Lightning Field* will have encountered it through them first. It has been argued that they create such a strong sense of expectation that they risk disappointing the visitor who arrives at the site—especially since this usually happens at mid-day, a time when the poles are barely visible and the site refuses to match any of the impressive images that are likely to have lured one there (Beardsley 1981, 36–37). If the artist's manifold attempts to control the perception of his work are indeed so central to it that it makes perfect sense to read them conceptually—as a part of the work rather than something that the sophisticated beholder should resist or circumvent—his control of *images* is an issue with far-reaching consequences. It has two main components: restricting official documentation to an extremely small number of photographs (all of which stem from Cliett's one visit and thus have a strikingly homogenous aesthetic), and rigorously exploiting their copyright to closely monitor and regulate the venues of their circulation.[13] Needless to say, public perception

13 De Maria's *Artforum* publication is a good example of these restrictive image politics. Beardsley reports being told by a representative of the Dia Foundation (the institution responsible for maintaining *The Lightning Field* and organizing the visits) that "it was

of the work has been profoundly shaped by these image politics, and while private documentation is a different issue altogether, it is worth noting that De Maria interferes with it as well by requesting his visitors to sign an agreement that commits them to not take any photographs of his work—with the result that those who do not comply (which I suspect to be a fair number) feel that they are breaking a personal contract with the artist.

All of this adds to the power of Cliett's photographs, which was put to an interesting test when the art magazine *Cabinet 3* asked a number of artists to "Please draw that famous photograph of *The Lightning Field* from memory"—a request that "plays on both the omnipresence of the photographs of the work *and* the fact that no one is allowed to take photographs of the work itself, and must thus rely either on published photographs or memory to reconstruct the work" (McFadden 2009, Fn. 45). It has been argued that Cliett's pictures do not simply reiterate the common iconography of the Western landscape; that they "both resonate with a sense of the American West and veer from our cultural understanding of that space—avoiding the aestheticized images of landscape photography and the cinematic expanses of the Western" (McFadden 2009, 84); that works like *The Lightning Field* "have an ambiguous relation to the land they occupy" in the sense that they are "dramatically sited" without being "involved with 'landscape' in any pictorial sense" as their sites are not "conventionally scenic" and their spaces "rather neutral, although very vast" (Baker 1983, 75). Yet no matter how conventionally un-scenic and presumably neutral the setting, and how atypical and veering its documentation, the absence or instability of an ideologically charged and concretized landscape that is evoked by the work and celebrated in both of these remarks has a single source: it draws from the notorious ambiguity of the 'complex pastoral' that is deeply engrained into the spatial design of the work itself.

I want to exemplify this point by reading *The Lightning Field* in terms of a 'middle landscape,' that pastoral ideal of combining the best of art with the best of nature (see Marx 1964, 160). As I have elaborated earlier, De Maria's signature work thrives on a carefully crafted tension between the natural site and the industrially produced field of steel poles installed in it—a tension that is organic (the poles are in the sense that the installation is part of the landscape) rather than dystopic (in the sense that it is brutally imposed on it and threatens to destroy its natural beauty). That the posi-

doubtful that De Maria would want to put his stamp of approval" on his highly critical essay by supplying photographs for it (1981, 37).

Fig. 3: The Lightning Field *during sunset.*

Source: Beardsley, John. (1984). *Earthworks and Beyond: Contemporary Art in the Landscape.* New York: Abbeville Press, 60.

tion of the poles has been determined by the logic of the grid makes the work reminiscent of cartographic appropriation, inviting the visitor to counter the vastness of the space around her by means of actively measuring and dissecting it: by walking through and around the field, estimating distances between poles and extending the grid across the entire plain, across the mountains in the background, across the continent. An imaginary engine of land colonialization, the grid toys with the possibility of transforming a wilderness into a garden.

For anyone who has seen the photographs depicting the artwork during a thunderstorm, the site resonates with the destructive power of the thunderstorms that it has been designed to attract, and whose dramatic imagery is most likely evoked at the sight/site of the field. Yet powerful as these associations are, the common mode of in situ perception is not the confrontation with the destructive *power* of nature but rather with its *obscurity*—the experience of vastness and infinity no less profound but entirely different in its perceptive parameters. It takes time and demands, first and foremost (and not unlike Henry Thoreau at Walden Pond, to whom I am about to turn) the beholder's patient investment.[14] And while experiencing the work on site is by no means incomplete without a thunderstorm—"The light is as important as the lightning" (De Maria 2010, 120)—the greatest challenge that the work has in store for its visitors is the fact that most of the time that one spends there it is barely visible. Only at short periods during dusk and dawn the poles stand out from the landscape as they do in the famous pictures (fig. 3). And yet there is always the expectation that a spectacle of light (or something else) will happen; in De Maria's auguring words, "The sum of the facts does not constitute the work or determine its aesthetics. [...] The invisible is real" (2010, 119–120).

That one has come to this place, which is far away from just about everywhere, to experience this 'something' is likely to suspend one's usual mode of perception and replace it with a heightened state of awareness for things that usually go unnoticed: the slowly changing light, the various surfaces and textures of the scenery, the subtle smells and sounds (much of this is also at stake when visiting *Spiral Jetty*). If works of land art are usually categorized as sculpture, *The Lightning Field* may be better described as an installation, that stage-like and highly participatory form of art that is only fully realized when experienced physically, spatially, and over an extended stretch of time.[15] Yet it is not only the time spent on location—a strictly fixed period of about twenty hours—that matters in this regard. Spatio-temporal investment begins with the journey that the visitor has to make in order to get here; a journey consisting of extended periods of low visual stimulus when driving through the desert. Eventually, this journey will take one to Quemado, a small cattle town in the scarcely populated western part of New Mexico, where the visitor leaves her car and signs a liability waiver (which is, more than anything, a performative ritual catering to the artifi-

14 In his compelling reading of *The Lightning Field* as an emblem of sublimity, Beardsley stresses the two above-named aspects—power and obscurity—as two of its primary features; others are "the privations of solitude and silence," its vastness and "inference of infinity," and the fact that the poles "stand in stately succession, uniform in height and the distance between them. As they diminish in the distance, they create the illusion—like telephone poles or railroad tracks—of endless progression" (1984, 62–63).

15 In Baker's fitting words: "De Maria tends toward dematerialization in favor of heightened mental/psychological content" (1983, 83). He worked "with a minimum of physical transformation, instead superimposing a permeable scheme" (ibid.). For insightful discussions of installation art see Rebentisch 2003; Bruno 2002.

cially heightened state of awareness mentioned above) to continue the journey in a pick-up truck driven by a member of the site's maintenance staff (no artsy people but real cowboys). After another forty-five-minute drive into what feels like the remotest part of the state one is dropped off on site to be retrieved the next day. Which is to say that visiting the site is a highly elitist, time- and money-consuming enterprise even for the connoisseur of this kind of art. This exclusiveness is further enhanced by an intricate process of scheduling one's visit months in advance, limiting the number of visitors to a maximum of six, and requiring them to stay for an entire day to make sure that a sufficient dose of duration and isolation are part of the experience.

The high investment in the experience that one is bound to have creates a set of receptive conditions (ranging from sheer expectation to an awareness of one's body after the long journey and the self-consciousness that comes along with being on site among strangers) that is just as crucial for one's perception of the work as its mediation through the documentary material that one has encountered before getting there. Yet once on site, something other than the obscure field of poles or the images that one has in mind assumes a primary importance: the simple but cozy cabin with three bedrooms, a fireplace and a richly stocked refrigerator. A former dwelling of a homesteader, abandoned a long time ago and only slightly altered for its new purposes, the cabin is briefly mentioned among the 'facts' given in De Maria's essay, omitted from Cliett's photographs, and yet highly significant for the work as a whole. In fact, the cabin makes possible the required duration of the stay on site in this hostile terrain. It is through the cabin that the place becomes inhabitable, and even though most visitors experience this aspect of the site only for a day, I am convinced that the sense of place that dwelling there creates—and that has been forced upon the visitors by the special regulations that everyone has to follow—is formative for the experience that one makes. I would indeed go so far as to claim that, in conjunction with the field of poles itself, the cabin marks the symbolic center of the topography in which the work is located. It turns *The Lightning Field* into a 'middle landscape'—and thus makes it distinctly pastoral What Leo Marx wrote about Thoreau's *Walden* in this regard thus turns out to be strikingly fitting for *The Lightning Field* as well, which is why I want to quote him at some length:

> The seemingly realistic setting may not be a land of fantasy like Arcadia, yet neither is it Massachusetts. On inspection it proves to be another embodiment of the

American moral geography—a native blend of myth and reality. The hut beside the pond stands at the center of a symbolic landscape in which the village of Concord appears on one side and a vast reach of unmodified nature on the other. As if no organized society exists to the west, the mysterious, untrammeled, primal world seems to begin at the village limits. As in most American fables, the wilderness is an indispensible feature of this terrain, and the hero's initial recoil from everyday life carries him [sic] to the verge of anarchic primitivism. "We need the tonic of wilderness," Thoreau explains, using the word "pasture" to encompass wild nature: "We need to witness our own limits transgressed, and some life pasturing freely where we never wander." (The combined influence of geography and the romantic idea of nature—sublime Nature—gives rise to attitudes held by a long line of American literary heroes from Natty Bumppo to Ike McCaslin.) But Thoreau is not a primitivist. True, he implies that he would have no difficulty choosing between Concord and the wilderness. What really engages him, however, is the possibility of avoiding that choice. [...] In *Walden*, accordingly, he keeps our attention focused on the middle ground [or middle landscape] where he builds a house, raises beans, reads the *Iliad*, and searches the depths of the pond. (Marx 1964, 245–246)

Yet despite all of the obvious similarities, there are important differences that need to be noted. Whereas Thoreau retreats to Walden Pond to recuperate his own senses and document an experience that included all of the above-named things, De Maria sends art-hungry travelers (the "heroes" of his topography) to New Mexico's high desert, an enterprise for which he recycles an abandoned hut and builds his laboratory of the senses, a field of industrially produced steel poles, next to it. The rationale of this latter endeavor is to raise a 'crop' that (besides the gain of prestige in the art world) may best be described as an aesthetic experience; a crop that draws together the site of production (the garden) and marketplace to which Thoreau still takes his goods (beans as well as books). In both cases, however—and this similarity is important here—the focus of the action is on cultivating a middle landscape with the idea of harvesting a perfect mix of art and nature. Thoreau accommodates himself with the machine's ambiguous presence in the garden by way of creating a tacit record of it, for example in the famous passage of the railroad both disrupting the silence at the pond and announcing the arrival of "many restless city merchants" (*Walden* quoted in Marx 1964, 250) with whom Thoreau traded part of his crop.

De Maria intensifies the ambiguity of the machine by *creating* rather than recording one and deliberately *putting* it into the garden. Akin to Thoreau's critique of consumerism, *The Lightning Field* was to challenge the commodification of art on the rise at the time by exploiting nature in ways that made the artwork practically unsellable. In this important sense, *The Lightning Field*'s primary raison-d'être is, just like that of Thoreau's *Walden*, to cultivate a heightened sense of awareness for one's natural environment. Yet it was also designed to unsettle the romantic cult of the artist that transcendentalists like Thoreau still fully embraced by blurring the line between artistic and industrial production. In taking the de-romanticization of artistic practice that had become a core concern in the postwar American art scene from the confines of the studio to the American West, he turned the figure of the artist as machine in the studio into a veritable machine in the garden.[16] In De Maria's industrial pastoral the ambiguity of the machine resides in both the work and its creator, whose ubiquitous presence haunts his work through the manifold attempts to execute control over its visitors' experience.

But De Maria's investment into that middle landscape comes at a higher price than having to deal with the artist's haunting presence. It reiterates the old myth of the empty continent (neither the significance that this region has for its indigenous population nor the nearby presence of one of the world's largest radio telescopes are ever mentioned in the official texts); it romanticizes loneliness and isolation, which are quintessential for having the profound experience that he is after; and it elevates the site designed to generate this cherished experience in an almost spiritual manner, safe-guarding it like the "high priest" of a sacred shrine.[17] Recent scholarship has made a compelling case for reading this role not as a mere expression of the artist's bold infatuation with power and control but rather in terms of his consistency as a conceptual artist. Yet in the case of *The Lightning Field*, the conceptual dimension of the work consistently takes us back to the site itself. The artist's strict refusal to collaborate in the in-

16 Jones describes the postwar transformation of the American artist in her book *Machine in the Studio*. The first stage of the development moves from romantic ideas of self-expression embodied by the artists associated with Abstract Expressionism to industrial ideas of artistic production epitomized in Andy Warhol's factory. According to Jones, Smithson is at the forefront of a next phase that moves away from the studio.

17 *Time* magazine dubbed De Maria the "high priest of danger" in an article about the first showing of his *Bed of Spikes* from 1968/69, which already uses a grid of stainless steel poles with pointed ends, yet on a much smaller scale (so that visitors could actually fall into them). Thomas Kellein suggests that the poles of the *Bed of Spikes* toy with the oblique, a symbolism that he sees resurfacing in *The Lightning Field*, which he reads in dialogue with religious rituals of ancient Egypt (1988, 54–55).

terpretation of this work, the thorough control of its documentation and the resulting repetition of the same few familiar tropes and images, the elaborate regulation of the visit, the striking fact that *The Lightning Field*, in spite of its permanent exposure to the forces of nature, does not seem to 'age'—all of this pushes it to an idealized, mythical space outside of time to stabilize it there.

The *Spiral Jetty* as Entropic Paradies

Robert Smithson's *Spiral Jetty* could hardly be more different in these regards. Submerged under the Great Salt Lake's thick red water between 1971 and early 2002 and mostly invisible during this long period of time, it became an object both in the material and the pictorial world soon after the work had been completed. Ravishing photographs by Gianfranco Gorgoni, whom Smithson invited to the site (just like De Maria would invite Cliett seven years later), were circulating widely, and through their repeated reproduction in the art press and general media some of them quickly assumed an iconic status. Yet whereas De Maria responded to the necessity of photographic documentation by turning *The Lightning Field* into an apparently stable, rarified object located in a mystic place untouched by the passing of time, Gorgoni (under Smithson's direction) and Smithson (in the few publicized photographs made by the artist himself) aimed at creating a visual record that deliberately works against stabilizing either the site or the artwork occupying it.

Even though the pictures were taken only shortly after its completion, sedimentations of salt had already begun to mark the fringes of the sculpture. The photographs highlight and stage this aspect of the work as a way of pointing toward a natural process of aging that is enhanced where the sculpture wards off the water of the lake. Tilted angles, unusual perspectives and frames create the sense of a vibrant, pulsating motion (that many of the photographs were taken from a moving helicopter certainly adds to this effect).[18] In Gorgoni's most famous 'portrait shot,' which he took

18 The dynamic quality of the early pictures becomes especially palpable when comparing them to those in Cooke and Kelly's volume on the *Spiral Jetty*. Published shortly after the earthwork's resurfacing and most likely inspired by it, it contains a series of stunning photographs documenting this process. Different from Gorgoni's pictures, however,

when approaching the sculpture from the vast space of the lake at a relatively low height, it seems to roll down from the scenic elevation that marks Rozel Point, ballooning toward the beholder before curling back into the center of the picture plane (fig. 4). Traces of heavy machinery—the bulldozer and the trucks that were used to build it—are clearly visible (as they are on the earthwork itself to this day), referencing not only the immense effort of construction—it took 292 truck-hours to build the *Jetty* (Hobbs 1981, 191)—but also the heavy machinery employed in it. That similar dams are built for railroads intensifies this presence of the machine in the garden—a theme that is tacitly reiterated in the background where another, barely visible jetty, the lonely remnant of an unsuccessful attempt of drilling for oil, stretches out into the lake in a line too straight not to be man-made.[19] Other photographs move in more and more closely, pushing the spatial context out of the frame to let the dynamic shape of the earthwork take center stage (fig. 5), dropping down to the water's mirror-like surface to look outward from a position near the sculpture's center across three disconnected, curving rows of salt-encrusted rocks (fig. 6) to finally zoom in on the salt crystals 'growing' on these rocks in an extreme close-up shot that exposes the crystals' spiral shape (fig. 7).[20]

The visual record thus created depicts an object at once sturdy and unstable; an object that resists being 'centered' in a unifying view. In doing so, it engages its beholder in a highly dynamic and participatory process of 'scaling' which Smithson hoped would undergird the in situ experience of the earthwork itself:

The scale of the Spiral Jetty tends to fluctuate depending on where the viewer happens to be. Size determines the object, but scale determines art. A crack in the wall if viewed in terms of scale, not size, could be called the Grand Canyon. A

---

they are strikingly inclined to depicting the earthwork as unified rather than unstable, probably because they are enthralled by the 'miracle' of its regained visibility as a material object.

19 Smithson mentions his familiarity with Leo Marx's *The Machine in the Garden*, which had been published six years before the construction of the Spiral Jetty, in an interview with Moira Roth (Smithson 2004 [1973], 94).

20 When Smithson first published his "Spiral Jetty" essay in 1972 in a volume on *Art in the Environment* edited by Gyory Kepes, he illustrated it with nine of Gorgoni's photographs organized in a movement that successively closes in on the work in a "double-counter-clockwise rotation (once through the air, once along the surface of the Jetty)" that resembled an "almost Icarian fall from the general to the particular" (Roberts 2004, 130). Posthumous reprints of the essay—Smithson died in a plane-crash in 1972—often published the carefully sequenced photographs out of order.

room could be made to take on the immensity of the solar system. Scale depends on one's capacity to be conscious of the actualities of perception. When one refuses to release scale from size, one is left with an object or language that *appears* to be certain. For me, scale operates by uncertainty. To be in the scale of the Spiral Jetty is to be out of it. (1996 [1972], 147)

Linking the stable, disengaged notion of size to 'mere objects' and the dynamic, participatory notion of scale to 'art' must be read as a direct response to Fried's dismissal of minimal art as literalist. Smithson pushed his phenomenological politics of perception laid out in this passage even further by creating a work that spans across different media. *Spiral Jetty* is indeed not only the title of the earthwork located at Rozel Point, but also of a film and an essay about it (the quote above stems from the "Spiral Jetty" essay). And while the film goes furthest in realizing Smithson's artistic program of decentering his signature work (using the non-human camera-eye to derail anthropomorphic modes of perception in ways that he later mapped out in his essay "Atopic Cinema"), all three versions of the *Spiral Jetty* are artworks in their own right. Which means that they lend their own media-specificity to a conjoint and non-closural effort of artistic creation.

Again, this artistic program can be read in terms of transforming the figure of the artist from the postwar 'machine in the studio' (prototypically embodied by Andy Warhol and his factory) into the artist as 'machine in the garden': at least two of the three *Jetties*—the earthwork and the film—are immediate products of industrial production taken to a 'natural' scene, which they transform in order to harvest a certain kind of (aesthetic) experience. And yet it is even more important to note that the multiple *Jetties* demand of their beholders that they engage in acts of 'scaling' across different media: while the actualities of perception are still made tangible by engaging with art objects across space, these objects materialize in spaces made available by specific media. Many scholars have insightfully written about this complex constitution of the *Spiral Jetty*, usually in the spirit of

Figs. 4–5: Spiral Jetty *rendered as unstable object by Gianfranco Gorgoni for Robert Smithson.*

Source: Roberts, Jennifer L. (2004). *Mirror-Travels. Robert Smithson and History.* New Haven: Yale University Press, 131–134.

Figs. 6–7: Spiral Jetty *rendered as unstable object by Gianfranco Gorgoni for Robert Smithson.*

Source: Roberts, Jennifer L. (2004). *Mirror-Travels. Robert Smithson and History.* New Haven: Yale University Press, 131–134.

rendering it as a showcase of postmodern art.[21] And if the critical discourse about the work tended to neglect these multiple 'life-forms' of the artwork for the first three decades of its existence, a more adequate view of it has by now been firmly established.[22] Yet there is more than a postmodern quarrel with art's traditional 'objecthood' and the respective role of the artist at stake in the rejection of stability and closure that lies at the heart of the *Spiral Jetty* (or rather constitutes the empty center around which the work revolves). In the remainder of this essay, I will follow Jennifer Roberts in arguing that Smithson's striking commitment to the dynamic and processual indicates (and possibly even indexes) a concern with history that sets him apart from his postmodern colleagues.[23]

Smithson's concern with history is clearly inscribed into the site that he selected for his monumental earthwork. The last stop for water and refreshments en route to it is at Golden Spike Monument, which commemorates the 'wedding of the rails' completing the transcontinental railroad with daily reenactments. Even if one does not stop here, the monument is impossible to miss, not least since the gravel road announcing the passage from 'civilization' to 'remoteness' begins right where it stands.[24] The gravel road that one enters here takes the visitor into a stunningly vast valley and all the way down to Rozel Point, which lies a good 45 minutes down the road. But shortly before getting there, the visitor is confronted with even

21 See for example Owens 1979 and 1980; Krauss, 1977; Reynolds 2003; Jones 1996; Shapiro 1995. Roberts lucidly summarizes this scholarship as follows: "It has demonstrated the challenges that Smithson's work posed to the totalizing impulses associated with high modernism by detailing his unabashed contamination of idealistic aesthetics with materiality, contingency, and physical labor; his dislocation of the traditional sites of artistic production; his complication of the authorial function; and his violation of disciplinary boundaries" (2004, 6).

22 The publication of *The Spiral Jetty* (2005), edited by Cooke and Kelly, marked a giant step in establishing this view.

23 Among the many things included in the *Spiral Jetty* project file, Roberts found "maps of the prehistoric shoreline of what is now the Great Salt Lake, several books and brochures about the construction of the transcontinental railroad, and an article on *kutsavi*, a food traditionally prepared by the Ute Indians by drying the tiny brine shrimps found in the lake (it was akin, apparently, to bacon bits)"—explicit references that "help to differentiate Smithson's work from that of many of his more properly minimalist or conceptualist colleagues, who tend to shrink away from making overt historical gestures" (2004, 1).

24 One can thus argue that the site, vaguely defined as it is, begins there and includes the oil dwelling ruin; some of Smithson's drawings support this view. See Roberts 2004, 1–2.

more of the site's historicity as she passes through an area marked by a striking assemblage of relicts from the past (the ruins of the aforementioned industrial jetty, an old car wreck, remnants of drilling tools) spread out in the mud and sand bordering the shoreline. The historicity of this second 'gateway' (the first one being Golden Spike) is materialistic rather than symbolic, a gateway of ruination and junk that marks a shift from historical to entropic time.

The artist's "propensity for 'distressed' land areas rather than unsullied virgin land" (Baker 1983, 83) is well known in art circles. In fact, it was Smithson himself who made it an explicit part of it into his signature work when he chose to write in his "Spiral Jetty" essay about his excitement of finding the former oil drilling site close to the spot that he was bound to select:

> Two dilapidated shacks looked over a tired group of oil rigs. A series of seeps of heavy black oil more like asphalt occur just south of Rozel Point. For forty or more years people have tried to get oil out of this natural tar pool. Pumps coated with black stickiness rusted in the corrosive salt air. A hut mounted on pilings could have been the habitation of "the missing link." A great pleasure arose from seeing all those incoherent structures. This site gave evidence of a succession of man-made systems mired with abandoned hopes. (1996 [1972], 146)

Describing the (extended) site of the *Spiral Jetty* as an entropic wasteland is hardly coincidental. Smithson had expressed an overt distaste for Arcadian longings in an earlier piece of writing, the first of several essays in which he conceptualized his earthworks: "the 'pastoral' [...] is outmoded. The gardens of history are being replaced by sites of time" (1996 [1968], 105). Pastoral ideals persist where art, under the guise of modernism, lends itself to transforming "the 'gloomy' ruins of aristocracy [...] into the 'happy' ruins of the humanist. Could one say that art degenerates as it approaches the garden?" (ibid.). Of course, the question is highly rhetorical. "The 'garden traces'," Smithson goes on to argue, "seem part of time and not history, they seem to be involved in the dissolution of 'progress.' It was John Ruskin who spoke of the 'dreamful Hammers' of the geologist, as they destroyed the classical order. The landscape reels back into millions and millions of years of geological time" (ibid).[25] The move from 'gardens of

25 In a later essay (the last he wrote before his death) Smithson celebrated Frederick Law Olmsted, the landscape architect who built Central Park in the mid-nineteenth century, as 'America's first earthwork artist' (1996 [1973], 164). For Smithson, he was worthy of this title for transforming the industrial wasteland that had been there before into a

history' to 'sites of time' thus lays out the itinerary of an artistic vision that rejects the conjunction of modernism and humanism against which the avant-garde movements of the 1960s had turned to get straight to the geological formation of 'deep time,' a concept that has only recently taken full shape as a critical resource in the continuous attempts to get beyond enlightened formations of knowledge (see McGurl 2011).

In following Roberts's lucid account that Smithson embraced an idea of "history as a material residue, an ever-accumulating remainder of time" (2004, 5), it is important to stress that the implied temporality was, indeed, geological rather than historical. It was not so much steeped in Marxist ideas of 'historical materialism' that were circulating at the time as it was founded in physical science; more precisely, it was founded in the understanding of crystallography that Smithson had developed over the course of the 1960s and that led him to posit "history as an entropic process of material deposition that settles itself into a final crystalline stillness" (Roberts 2004, 4).[26] Doubling the earthwork's spiral shape on the microscopic level that Gorgoni so artfully captured in one of his pictures (fig. 7), the sedimentations of salt on the rocks that form the earthwork foreshadow and index precisely this entropic notion of history, and it is in this important and inherently dynamic sense that the salt of the lake is to the *Spiral Jetty* what light and lightning are to *The Lightning Field*—namely that dynamic element of the site that on which the artwork depends in quintessential ways.

Approached this way, however, Smithson appears not so much as the alleged textbook-postmodernist as whom he is usually treated, but rather as an 'exomodernist' before his time—a strategic position that Mark McGurl evocatively describes as "*outside of* rather than *after* the modern and the postmodern, displacing postmodernism's notorious pillaging of past historical styles by trying to imagine what lies beyond or alongside style (2011, 181, emphasis in the original). The exomodern commitment to a historical logic grounded in notions of space rather than time settles nicely with Smithson's above-mentioned plea to move from 'gardens of history' to 'sites of time' (1996 [1968], 105). But it also brings out a fundamental paradox at the heart of his signature work. For to fathom and claim a position

landscape that embraces and stages historical change rather than turning it into a picturesque surface.

26 Roberts delineates this development and its impact on Smithson's work in chapter 2 of *Mirror-Travels* (2004).

beyond rather than *post-style* meant that his conception of the *Spiral Jetty* would have to partly hold on to the very pastoralism that he intellectually rejected; that he would have to reverse it rather than leaving it behind. For conceptually, the *Spiral Jetty* is still very much a garden—it is an artificially transformed and cultivated plot of land. Yet in this garden growth leads to stillness and extinction rather than harvest and progress. This has the effect of putting on its head the idea of cultivation that art, cannot do without, simply because it always entails some sort of ideal of productive transformation. Ever since the modern redefinition of art, these ideals revolve around fostering a heightened aesthetic sensibility (see Rancière 2004). And even when understood with a keen eye on aesthetic politics, as "ways of doing and making" that intervene in the general scheme of distribution of the sensible with substantial impact on what is visible, sayable and audible at a given point in time, artistic practices remain indebted to "the formation and education of a specific type of humanity" (ibid. 24).

The *Spiral Jetty* does not operate outside of this logic; in fact, it forms and educates by virtue of its reversed pastoralism. As an entropic garden in which inorganic matter is to deposit until it has reached a final point of perfect stillness, it is invested in an ethics that seeks to cultivate 'better subjects' by confronting them with the relativity, or even insignificance of human existence on this planet. The existentialism that surfaces here was certainly not an intended part of Smithson's work. But it might help explain the peculiar "need to pre-figure a transcendent, eternal condition beyond the limitations of history" that Roberts, to her great surprise, found counterbalancing the artist's "convictions about the material persistence of history"—a condition that "was not to be metaphysical but rather, in his terms, 'infraphysical,' [...] located within rather than without the material world" (Roberts 2004, 9). For Smithson, this had to involve a "transcendental state of matter" (quoted in ibid.)—which turns his *Spiral Jetty* into a real entropic paradise.

Works Cited

Auping, Michael (1983). Earth Art: A Study in Ecological Politics. In Alan Sonfist (ed.). *Art in the Land: A Critical Anthology of Environmental Art*, 92–104. New York: E.P. Dutton.

Battcock, Gregory (ed.). (1973). *Idea Art: A Critical Anthology*. New York: E.P. Dutton.

Beardsley, John (1981). Art and Authoritarianism: Walter De Maria's *Lightning Field*. *October*, 16, 35–38.

— (1984). *Earthworks and Beyond: Contemporary Art in the Landscape*. New York: Abbeville Press.

Baker, Elisabeth C. (1983). Artworks on the Land. In Alan Sonfist (ed.), *Art in the Land: A Critical Anthology of Environmental Art*, 73–84. New York: E.P. Dutton.

Boetzkes, Amanda (2010). *The Ethics of Land Art*. Minneapolis: University of Minnesota Press.

Bruno, Giuliana (2002). *Atlas of Emotion: Journeys in Art, Architecture, and Film*. New York: Verso.

Casey, Edward (2004). Mapping the Earth in Works of Art. In Bruce V. Foltz and Robert Frodeman (eds.), *Rethinking Nature: Essays in Environmental Philosophy*, 260–270. Indianapolis: Indiana University Press.

Cooke, Lynne (2005). A Position of Elsewhere. In Lynne Cooke and Karen Kelly (eds.), *Robert Smithson: Spiral Jetty*, 53–71. Berkeley, CA: University of California Press.

— (2008). Preface. In Kenneth Baker, *The Lightning Field*. New Haven: Yale University Press.

De Maria, Walter (1988). On the Importance of Natural Disasters. In Thomas Kellein (ed.), *Walter De Maria: 5 Kontinente Skulptur*, 45. Staatsgalerie Stuttgart.

— (2010). Some Facts. In Simon Morley (ed.), *The Sublime*, 118–121. Cambridge, MA: MIT Press..

Fried, Michael (1995). Art and Objecthood. In Gregory Battcock (ed.), *Minimal Art: A Critical Anthology* 116–147. Berkeley, CA: University of California Press.

Graziani, Ron (2004). *Robert Smithson and the American Landscape*. Cambridge, UK: Cambridge University Press.

Greenberg, Clement (1955). "American-type" Painting. *Partisan Review*, 22.2, 179–196.

Hickey, Dave (1971). Earthscapes, Landwork, and Oz. *Art in America*, 59, 40–49.

Hobbs, Robert (1981). The Works. In Robert Hobbs (ed.), *Robert Smithson: Sculpture*. Ithaca, NY: Cornell University Press.

Jones, Caroline (1996). *Machine in the Studio: Constructing the Postwar American Artist*. Chicago: The University of Chicago Press.

Kellein, Thomas (1988). Die Kunst des Walter De Maria. In Thomas Kellein (ed.), *Walter De Maria: 5 Kontinente Skulptur*, 31–57. Staatsgalerie Stuttgart.

Krauss, Rosalind (1977). The Double Negative: A New Syntax for Sculpture. In Krauss, *Passages in Modern Sculpture*, 243–288. New York: Viking.

— (1979). Sculpture in the Expanded Field. *October*, 8, 30–44.

Kwon, Miwon (1997). One Place After Another: Notes on Site-Specificity. *October*, 80, 85–110.

Lippard, Lucy (1997). *Six Years: The De-materialization of the Art Object.* Berkeley: University of California Press.

Marx, Leo (1964). *The Machine in the Garden. Technology and the Pastoral Ideal in America.* New York: Oxford University Press.

— (1970). American Institutions and Ecological Ideas. *Science*, New Series, 170, Nov 27, 945–952.

— (1999/2000). Afterword: The Machine in the Garden. *The Massachusetts Review*, 40.4, 483–496.

— (2005) On Recovering the 'Ur-Theory' of American Studies. *American Literary History*, 17.1, 111–134.

McFadden, Jane (2009). Earthquakes, Photoworks, and Oz: Walter De Maria's Conceptual Art. *Art Journal*, 68.3, 68–87.

McGurl, Mark (2011). The New Cultural Geology. *Twentieth Century Literature*, 57.4, 380–390.

O'Doherty, Brian (1999). *Inside the White Cube: The Ideology of the Gallery Space.* Berkeley: University of California Press.

Owens, Craig (1979). Earthwords. *October*, 10, 120–130.

— (1980). The Allegorical Impulse: Toward a Theory of Postmodernism. *October*, 12, 67–86.

— (1992). *Beyond Recognition: Representation, Power, and Culture.* Berkeley: University of California Press.

Rancière, Jacques (2004). *The Politics of the Aesthetics: The Distribution of the Sensible.* Trans. Gabriel Rockhill. London: Continuum.

Rebentisch, Juliane (2003). *Ästhetik der Installation.* Frankfurt am Main: Suhrkamp.

Reynolds, Ann (2003). *Learning from New Jersey and Elsewhere.* Cambridge, MA: MIT Press.

Roberts, Jennifer L. (2004). *Mirror-Travels. Robert Smithson and History.* New Haven: Yale University Press.

Rosenberg, Harold (1972). *The De-definition of Art.* Chicago: University of Chicago Press.

Sandler, Irving (1988). *American Art of the 1960s.* New York: Harper and Row.

Shapiro, Gary (1995). *Earthwards: Robert Smithson and Art After Babel.* Berkeley: University of California Press.

Smithson, Robert (1996 [1968]). A Sedimentation of the Mind: Earth Projects. In Jack Flam (ed.), *Robert Smithson: Collected Writings*, 100–113. Berkeley: University of California Press.

— (1996 [1970]). Interview with Paul Toner. In Jack Flam (ed.), *Robert Smithson: Collected Writings*, 234–241. Berkeley: University of California Press.

— (1996 [1971]). Atopic Cinema. In Jack Flam (ed.), *Robert Smithson: Collected Writings*, 138–142. Berkeley: University of California Press.

— (1996 [1972]). The Spiral Jetty. In Jack Flam (ed.), *Robert Smithson: Collected Writings*, 143–153. Berkeley: University of California Press.

— (1996 [1973]). Frederik Law Olmsted and the Dialectical Landscape. In Jack Flam (ed.), *Robert Smithson: Collected Writings*, 157–174. Berkeley: University of California Press.

— (2004 [1973]). Interview with Moria Roth. In Eugene Tsai, *Robert Smithson*, 80–95. Los Angeles: Museum of Contemporary Art and University of California Press.

Sontag, Susan (1966). *Against Interpretation and Other Essays.* New York: Farrar, Straus and Giroux.

Szeemann, Harald (ed.). (1999). *Walter de Maria: The 2000 Sculpture*, Exhib. Cat., Zürich: Kunsthaus Zürich.

Tiberghien, Gilles A. (1993). *Land Art.* Paris: Édition Carré.

The Machine in the Indigenous Garden: Leslie Marmon Silko's *Gardens in the Dunes*

Christa Grewe-Volpp

Few texts have had such an impact on American studies as Leo Marx's seminal book *The Machine in the Garden* from 1964. In it, he neatly developed the idea that the meaning of America has been defined by pastoralism ever since the age of discovery and that the new continent invited the utopian dream to actually realize a poetic fantasy (Marx 2000, 3). The garden came to be seen as an ideal middle landscape between a howling wilderness and the constraints of civilization, providing an artful balance between nature and culture. By the end of the eighteenth century, the pastoral idea of America had become manifest in the national imaginary, as expressed in Jeffersonian agrarianism and de Crèvecoeur's visions of the American farmer, according to which the cultivation of land nurtured a morality far superior to the corruptions of industrial England. Marx argued with Henry Nash Smith that the garden had become a master symbol of America; Americans "actually saw themselves creating a society in the image of a garden" (ibid., 143). The nineteenth century, however, disrupted the (ostensibly) harmonious relationship between man and nature when the machine, most notably the railroad, entered the Edenic landscape. Although some believed that the machine would serve the purposes of nature and enhance its properties and that it was even necessary to fulfill democratic ideals by releasing man from drudgery, others were more skeptical and foresaw the destruction of an idyllic continent with a concomitant loss of spiritual and moral values. According to Marx, these two conflicting attitudes towards technological progress have consolidated a major cultural divide in America still prevalent today, the dominant political and economic elite posited as affirmative of progress on the one hand, and "the adversarial viewpoint of a minority of political radicals, writers, artists, clergymen, and independent intellectuals" located on the other (ibid., 383). Those opposed included environmentalists and other thinkers critical of modernity, for whom the green landscape figured as a place of resistance against the technological and ideological forces of modernity.[1]

Marx's thesis has come under attack as representative of the Myth and Symbol School, which allegedly reduced the complex historical development of the American nation to a single idea. Others, like Richard Sennett in a review of the republished version of *The Machine in the Garden* in 2000, argued that it is "a timely book," "a cultural investigation" which still explains major aspects of American society (20). In this article, I take the validity of Marx's thesis in general for granted, but I do want to ask in which ways it might be incomplete in the context of the early twenty-first century. Marx himself admitted in the afterword to the new edition that "[t]oday [...] it would not be possible for me to write certain sentences in *The Machine in the Garden* that tacitly generalize about the thought and behavior of 'Americans' " (Marx 2000, 382). Since the 1960s, American Studies have, above all, explored the multiplicity and diversity of the American continent and have acknowledged the experience of its citizens "determined by their class, ethnicity, race, gender, or sexual preference" (ibid., 382). Today one cannot ignore the fact that the concept of the American pastoral as it was so aptly identified and analyzed by Marx was from its inception a biased idea of bourgeois white men which supported hegemonic power structures. Not only did the gendered view of nature (and implicitly of the garden) allow for the exploitation of the land as well as of women, as Annette Kolodny elaborated in her influential book *The Lay of the Land* (1975), the idea of America as a pastoral idyll also disregarded the fact that the taming of the wilderness brought with it the displacement and outright genocide of Native Americans.[2] One could even argue that the pastoral was imagined in opposition to the 'Indians' and everything they supposedly stood for, most notably their 'naturalness' and

1 It can be argued that the healing of the split between nature and culture has been the dominant concern of ecocritics, who have revealed how the dichotomy inherent in the American imaginary has resulted in the destruction of a natural environment on a particularly large scale on the North American continent. In the last decade new developments in ecocriticism, particularly New Materialism, have shown that notions of imbrication, interdependence and intra-activity must replace binary thinking and thus, implicitly, modify Marx's thesis. I am thinking particularly of Karen Barad's seminal study *Meeting the Universe Halfway* and Jane Bennett's *Vibrant Matter*, which have been highly influential for ecocritical thought. But that would be the topic of another paper.

2 In the following I use the terms "Native American" and "indigenous population" synonymously. "Indian" as an expression of colonization is marked with inverted commas. Most Native Americans identify themselves by their tribal affiliation.

'inability' to cultivate a vast continent. 'Indians' were not regarded as human beings equal to 'civilized' Europeans. The first Elizabethan settlers identified them with primal nature, which was sometimes seen as a hellish realm of darkness harboring ferocious wild people. However, often the Natives were idealized as noble savages living in harmony with Eden. As Marx noted, "They fit perfectly into the picture of America as a mere landscape, remote and unspoiled, and a possible setting for a pastoral retreat" (Marx 2000, 36). About a hundred years later, Robert Beverley still admired the 'Indians' for their gentle, joyous and simple ways which he attributed to their close affinity with a primordial garden. At the same time though he saw a detrimental influence of the lush land on the Englishman who succumbed to sloth and other vices (see Marx 2000, 85–87). In the eighteenth-century paradigm, Englishmen had to cultivate wild nature and thus create a "real" garden (a farm, a park), whereas the mythical garden with its noble savages remained a mere ideal with disastrous effects for Native Americans, who, by the end of the nineteenth century, were finally defeated in their claims for territory in the Battle of Wounded Knee. The then prevalent idea that the 'Indian' was doomed to extinction can be regarded as a form of guilt management as well as nostalgia for a 'vanishing race' and a pastoral garden.

Interestingly, one of the most renowned Native American writers of the late twentieth and early twenty-first century, Leslie Marmon Silko, picked up the American master symbol of the garden in her novel *Gardens in the Dunes* (1999) and made it the focal topic in her exploration of colonialist, capitalist ideologies and their effects on Native Americans at the turn of the 20th century. According to Hartwig Isernhagen, this should not be a surprise as "it was to be expected that a native writer would come and (re)claim the garden as cultural icon and emblem for indigenous writing" (2002, 176–177). In the following, I will analyze whether Silko's fictional creation of an indigenous garden corresponds to Marx's definition of a pastoral landscape and whether it too, functions as a retreat from the constraints and corruptions of civilization. I will argue that Silko does not fall back on idealized notions of the Native living in close harmony with a primordial nature. The fictional home of a vanishing desert tribe in the sand dunes along the Colorado River functions as a place of political resistance against the encroachments of a destructive culture. After all Silko emphasized in an interview with Ellen Arnold that "you just can't write an apolitical novel about gardens" (Silko quoted in Arnold 2000, 181)! I will point out what kind of machines intrude into her garden, and I will show that the Native garden and technology (which Marx identified as a counterforce to the idyllic version of the landscape (2000, 25, 26)) are not necessarily diametrically opposed to each other, but only in the context of colonialism. If Marx argues that the pastoral idea has defined the meaning of America ever since the age of discovery (ibid., 3), Silko redefines its meaning.

Gardens in the Dunes is the story of two young Sand Lizard girls who are expelled from their home in the southwest desert and exposed to the imperial and colonial practices at the end of the nineteenth century. The plot is thus set in a time when the Western frontier was officially declared closed and the remaining Native population removed to reservations. Many children were coerced into attending Indian boarding schools to be assimilated into mainstream American culture. Boarding schools, according to the *Harvard Encyclopedia of American Ethnic Groups*, "were considered the chief tool for achieving assimilation." Children were separated from their parents, forced to speak only English and learn 'civilized' ways of life. Richard H. Pratt, the founder of the (in)famous Carlisle Indian School in Pennsylvania, called the breaking up of "'tribal and race clannishness [...] a most important victory in getting the Indian toward real citizenship'" (117, 118). It was the time when the General Allotment Act (or Dawes Act) from 1887 was enforced as a result of which Native Americans lost about 60 percent of their tribal land (Ryan 2007, 129). Indigo, the younger of the girls, is captured by the Indian police and sent to a boarding school in California, from which she manages to escape. She ends up with a hapless couple, Edward and Hattie Palmer, who take her on a Grand Tour to the East Coast and to Europe, to England, Italy and Corsica. On her journey, she becomes acquainted with a variety of gardens which, as Angelika Köhler has noted, "function as images of historical paradigms, of cultural systems" (2002, 239); they allow Silko to explore ancient and contemporary ideologies in accordance with or contrary to the indigenous gardens in Arizona. Indigo's older sister, Sister Salt, stays closer to home. She ends up working as a voluntary prostitute and washerwoman at a construction site in California, where a huge dam is being built to provide cities like Los Angeles and agricultural plantations with water. It is an enormous capitalist enterprise which destroys the natural environment and fosters social exploitation and greed. Both sisters succeed in reuniting in the gardens in the dunes and are able to live again in line with their traditional values, at least for the time being. The different forms of land use with their implicit ideologies are consistently reflected through the eyes of the two sisters, mostly Indigo's. Her point of view serves as a narrative strategy that "reverses the ethnographic gaze" (Isernhagen 2003, 129) by removing the white man from the garden and rewriting the myth from an indigenous, female perspective. With the juxtaposition of the gardens and different methods of horticulture, Silko articulates her political vision, namely that the vacuity and destructiveness of colonial and capitalist enterprises, which include the exploitation and extinction of indigenous cultures, is opposed to a spiritual way of life based on traditional values of the sacredness of life and of sharing. In order to analyze this indigenous perspective, I will take a close look at representations of the gardens in the dunes and identify their political implications, especially in contrast to the Masque of the Blue Garden on Long Island (an artifice created by Edward's sister and her husband) as well as to Edward's own colonialist practices as a plant collector in Brazil and Corsica. I will also briefly look at representations of two gardens in Europe, one in England and the other in Italy, because both serve as alternatives to destructive Euro-American land use. It is the aim of this essay to point out how the novel employs the master symbol of the garden as identified by Marx to problematize ideas of identity which are connected to specific attitudes towards place.

Gardens in the Dunes begins and ends at the indigenous gardens in the desert, thus providing a cyclical and geographical frame for a novel that includes a large number of multiethnic characters and an equally large number of sites across the USA and Europe. It is the home of the Sand Lizards, a fictional tribe closely resembling long extinct tribes who lived in some side canyons of the Colorado River, and were "totally wiped out" by gold miners and ranchers (Silko quoted in Arnold 2000, 163). Silko created four women as the last survivors of the Sand Lizards who embody and are fortified by their traditional agricultural and spiritual ways of life: Grandma Fleet, Sister Salt, Indigo and their mother. What characterizes them is their difference even among other Native American people, "but they were proud to be known for their contrary ways" (48).[3] They fiercely protected their autonomy and hid from white colonizers in places hard to access. At the same time, they welcomed friendly outsiders into their tribe. "Sex with strangers was valued for alliances and friendships that might be made." No

3 Quotations directly taken from the novel will be indicated only by the page numbers in the following text.

matter what skin color the baby had, "the Sand Lizard mother's body changed everything to Sand Lizard inside her" (202). Should they get into a fight with their enemies, they never took prisoners, because "Sand Lizards didn't have much use for slaves" (49). Although their home in the desert is a small local place, it is not a niche cut off from the rest of the world. On the contrary, the cultural practices of the tribe create a sense of ethnicity that is radically inclusionary, as long as its people are allowed to pursue their peaceful, self-sustaining life. In their community, race is not a marker of identity, nor is national belonging (other than in the ideal Euro-American pastoral in which the garden was the domain of white American men). According to Silko, it is the destroyers and exploiters who draw boundary lines (Silko quoted in Arnold 2000, 170), who cannot tolerate the existence of native groups and are responsible for the disappearance of the majority of the indigenous population. The creation of the fictional Sand Lizards allows Silko to write against notions of Native American tribes as distant, closed cultural groups or as victims. Her four women are strong, self-reliant, intelligent, open-minded and spiritually secure, even if they are finally overwhelmed by historical forces.

What makes them so strong is their deep connection to place, in particular to their gardens. Their feeling of belonging is not the result of taming the wilderness, a concept they are not even familiar with, nor can it be associated with frontier thinking and inherent ideas of Manifest Destiny. The tribe belongs because their settling the sand dunes dates back to mythical times when their ancestors were invited by the Sand Lizard to stay and to cultivate seeds. Furthermore, belonging is not linked to possession either, even if the people use the earth for agricultural purposes, because the land is not theirs in the Western legal sense. The most important requirement for staying in the gardens is the principle of exchange, which includes the community of all elements of the environment. It implies leaving seeds for the ancestors, for birds and animals, for bees and insects. This way, the plants reseed themselves (15), and the garden grows even if humans cannot tend to it for whatever reasons. Not only is this form of agriculture a smart method to ensure sustainability in a place deemed hostile to human survival by white settlers, it also nurtures its people spiritually. Plants and animals are "kin" to Native Americans. Taught by Grandma Fleet about the "decorum of gardening" (Fischer-Hornung 2009, 321), Indigo learns to respect plants, to speak to them, to consider them as valuable living beings that will feed her and please her aesthetically. Human beings, plants and animals belong to the great cycle of life. When Grandma Fleet prepares to die, she tells her grandchildren to bury her underneath the apricot tree so that her body can be fertilizer for another life form, thus partaking in the principle of exchange even after death.

Similar to the American pastoral in Jeffersonian agrarianism, gardening creates moral values and brings the gardener closer to God, but unlike agriculture in the white American pastoral, indigenous gardening does not subordinate an inanimate nature, nor does it bring forth independent Jeffersonian husbandmen. Moral values imply kinship with the natural environment, a respect for the community of humans and nonhumans, radical openness to newcomers, altruistic behavior and pacifism. New seeds are always welcome; they enhance the diversity and therefore, the richness of the place, which is thus paradoxically local and global at the same time. Cultivation is not the sole task of man (or woman), but of the earth itself, which is, as new materialists would argue, an active agent in the creation of the world. Horticulture, especially cross-pollination and hybridization—all of which Indigo learns from Grandma Fleet—is valued not for its potential for profit but as a means to grow food and to participate in the processes of life. Native gardeners also alter nature, but they do it much more gently than white immigrants, enhancing, not destroying what they found to grow wherever they settled. As Dorothea Fischer-Hornung argues, horticulture thus becomes "the metonymic core of Silko's exploration of what it means to be civilized" (2009, 313). Indigenous people like Indigo have created their own forms of civilization. They are not one with the wilderness, "remote and unspoiled," leaving no marks, as white stereotypes would have it. Such an attitude "demeans Indians: It makes them seem simply like an animal species, and thus deprives them of culture" (Cronon 1986, 20).

Another aspect that characterizes the indigenous garden is its genderedness. Of course in the American pastoral, the garden was also gendered female. But it implied the domination and exploitation of the land, the rape of "mother earth" and the simultaneous subordination of women on the North American continent during the time of its settlement (see Kolodny 1975). Silko's gardens in the dunes are a feminized space too, but this space serves as a concept of female empowerment. Not only are the Sand Lizards a matriarchal tribe, celebrating fertility and communality, even their mythology is based on a female figure, the Sand Lizard. Their matriarchal society however, does not exclude or dominate men or male aspects of life, but rather includes them as necessary parts of a whole.[4] Not only do they welcome different men, they also revere Grandfather Snake as an important figure in their cosmos and as part of the gardens. Unlike in the Christian origin story, the snake does not symbolize evil and initiate the expulsion of Adam and Eve from Paradise. Here the snake is a protector, a sacred animal ensuring the continuance of life. Likewise, it is not a woman who is responsible for the fall; responsible are, as will be shown later, the political powers of colonialism and capitalism combined.

Both the Euro-American garden as pastoral and the indigenous gardens function as places of refuge. In the idealized version of Western culture, the garden allows a flight from the corrupt city into unspoiled nature. This flight, following a pattern going back to Virgil, implies a "contrast between two worlds, one identified with rural peace and simplicity, the other with urban power and sophistication" (Marx 2000, 19), or from the nineteenth century onwards, with industrialization and machine technology. The contrast creates a tension that can either be ignored in a simple escape from civilization, or it can make the pastoralist aware of the deficiencies of his urban world. The green world then enacts the desire to change the shortcomings of modernity (see Marx 1986). The pastoral as a mode—"because it speaks of a way of being-in-the-world" (Garber 1988, 432)—expresses a feeling which oscillates between a longing for peace and harmony on the one hand and for an improved civilized world on the other.

The dichotomy between a peaceful, simple green world and the destructive practices of Western civilization is also a dominant theme in *Gardens in the Dunes*. At first sight, one might even suspect that Silko follows what Marx called a "pastoral of sentiment" (Marx 1986, 5) associated with escapism. The Sand Lizard gardens are, after all, depicted as a true idyll, where the women live a sensuous happy life, nurtured by the cornucopia of a generous feminine nature, away from the dangers of the modern world. Does this mean that Silko propagates a romanticized version of primitivism? The answer must clearly be "no" for several reasons. For one, the dichotomy between the rural and the urban, between nature and civilization, does not exist within their culture but is enforced on the Native population by an external foreign colonialist power which is intent on destroying the indigenous culture, either by physical violence or by assimilation. The soldiers have already intruded into the gardens before but

4 Deborah A. Miranda (2007) explores the balancing of the male and female energies in Silko's novel.

mostly stay away because the dunes are hard to reach on horseback. The "aliens" brought "bloodshed and cruelty," "hunger and suffering," "disease and fever" (15). Under these extreme circumstances, the gardens became a "sanctuary" (15) providing protection from persecution and extinction and thus served much more literally as a direct political place of refuge than the pastoral green world of Euro-American culture has ever done. The indigenous gardens also do not exist in opposition to civilization *per se*, they *are* a civilized place with their own sets of rules and etiquettes, which are passed down from one generation to the next. And last but not least, the Sand Lizards are not torn between two ways of life; one a utopian ideal, the other a genuine possibility. They can only survive in the sand dunes because the colonialists are intent on their extinction in the world outside. 'Escape' thus assumes a totally different meaning in the indigenous context of the novel.

Silko juxtaposes forms of indigenous gardening with non-Native methods of land use and explores their discrepancies and underlying ideologies. An analysis of Euro-American appropriations of land reveals the degeneration of the pastoral ideal and points out its hidden agenda of racism. The most obvious examples in the novel are the Masque of the Blue Garden, Edward Parker's trips to Brazil as a collector of rare plants as well as the building of a dam in Arizona.

The Blue Garden is a creation of Edward Palmer's sister Susan James. Her husband Colin is one of the Robber Barons who accumulated massive wealth in the late nineteenth century with criminal-like economic methods including the exploitation of geographical and human resources, made possible in part by the building of the railroad and the easy accessibility to new, 'unsettled' land. Their grand mansion on Long Island and their elaborate Renaissance garden are prominent examples of conspicuous consumption, which, as American sociologist and economist Thorstein Veblen pointed out in *The Theory of the Leisure Class* in 1899, was meant to exhibit the wealth and social status of their owners. Susan does not value the old trees and bushes but is intent on replanting the property to turn it into an English landscape garden, which demands the uprooting of mature trees and other plants or, as Edward observes, the "demolition" of a beautiful park (161). Her new garden is to contain blue and white flowers merely for decorative effect. Not only are old trees uprooted, but other plants are forced to bloom out of season, "white tulips and white freesias carefully prepared by weeks in the cold room of the greenhouse" (195). Susan wants to impress her illustrious guests (including a bishop) at the Masque of the Blue Garden, an annual party founded to collect aid for the poor, but actually it gives her an opportunity to show off her and her husband's status and to be the center of attention herself. The Blue Garden serves an economic function by stabilizing power relations; there is constant talk about money and profit and creditors and the stock market in the James family. Nature is only a prop, abused and violated. Plants are dislocated and "forced," all for the materialistic purposes of capitalists. The indigenous population is equally abused because the appropriation of the land by Euro-Americans necessitated the displacement of its first people. On her forays into the neighborhood, Indigo comes upon the remnants of a tribe who used to live on all of what is now Long Island. They have one subsistence farm left to practice their own culture. The Blue Garden thus stands in stark contrast not only to the indigenous gardens in the dunes but is also a perversion of Jeffersonian ideals of agrarianism. Here no morally superior husbandman toils the earth, the Jameses are more likely personifications of the machine in the garden causing destruction and moral emptiness. Even Mr. Abbott's (Hattie's father's) ordinary garden cultivation, "limited to relief of hunger among the poor" (178), is actually just a philanthropic experiment which, in the final analysis, solidifies existing power structures. As Tamara Plakins Thornton observes, horticulture at the end of the nineteenth century was thought to teach "republican virtues: hard work, thrift, and the sacrosanct worth of private property" (quoted in Ryan 2007, 122). Whichever way you look at it: the creation of the 'white' gardens and the establishment of white hegemonic power rest upon the displacement of a native population be it people, animals or plants.

In the American pastoral, the symbol of the machine was surrounded by a rhetoric of progress (Marx 2000, 193). Technology and industrialization offered the opportunity of an improved civilized world for Euro-Americans, promising the pursuit of happiness and the liberation of the oppressed (ibid., 206). Although a minority of American men of letters expressed their doubts about the blessings of the machine, the majority regarded its use as their "birthright" (ibid., 205); it would help them shape a country where nature and culture existed side by side in a harmonious relationship. This imagined country was, of course, an idealized construction in which nature and 'natural' peoples were tamed (assimilated) or obliterated; by no means could they thrive on their own terms. Technology, industrialization and progress as phenomena of modernity, symbolized by the machine, produced very ambivalent reactions in the Anglo-European society, creating enormous seductive opportunities as well as feelings of alienation and social insecurity. But they were totally devastating for the survival of indigenous cultures. These new technologies did not function as a liberating force in the Native American gardens. One could even argue that whereas the ideology of the white pastoral was blind to the fate of the 'Indians,' the machine in the garden as technology and metaphor served as a tool to seal their effacement and to make new use of the land for economic profit. In the indigenous context, the machines are literally construction site equipment for the building of the dam, the railroad, technical instruments such as the camera, but also, on a metaphorical level, the practices of colonialism and capitalism. The railroad, for example, *the* master symbol of progress in the nineteenth century, takes Indigo away from Arizona and transports her to the East Coast and through Europe. It is thus not only an instrument of displacement, it also reverses the mythical journey West with its promises of individual fulfillment. Of course Indigo makes the best of her travels, as she connects with other people and other cultures, collects new seeds and ideas to 'hybridize' her indigenous way of life. Nevertheless the railroad distances her from the sand dunes. Her return west is finally not linked to a new beginning at an imagined frontier but to a movement home.

The mechanisms and effects of the 'machine' are highlighted quite clearly in Edward Parker, the stock figure of a colonialist at the end of the nineteenth century. He is a botanist who travels to far and exotic places to collect rare plant specimens, mostly orchids, in order sell them for substantial amounts of money on the American market.[5] Not only does he indiscriminately rob the native soil of its plants, he engages in criminal actions by smuggling roots and seeds out of the country. On his expedition to the Pará River in Brazil, however, he is outsmarted by even more unscrupulous smugglers who use him as a decoy in order to steal disease-resistant specimens of rubber tree seedlings for Kew Gardens in London to break Brazil's monopoly on rubber. They mercilessly leave him behind after setting fire to the jungle to dispose of any potential competitor, risking his death. Edward later travels to Corsica with Hattie and Indigo to steal citrus cuttings, also for the purpose of breaking the country's monopoly on citrus, which was much in demand around 1890. Most importantly,

5 For further information about the orchid mania of the nineteenth century and the early trafficking of plants see Dorothea Fischer-Hornung (2009).

his success would have solved his dire economic situation. Greed and social recognition are the driving force of his actions, which must be seen in the context of the political climate of his times, defined by capitalism and colonialism operating on a global scale. Silko points out the parallels between the destruction of whole ecosystems and their inhabitants not only on the North American continent, but also in South America and, centuries ago, in Europe. In *Gardens in the Dunes*, Indios in Brazil, for example, had to help collect rare orchids against their will and better judgment, otherwise they were "flogged or tortured" (133).

Edward's technical equipment reveals his attitude toward the natural environment and is consistent with nineteenth-century ideas of science and progress. In his laboratory/study in California, Hattie finds "lanterns, candles, tents, tarps, a folding shovel, a trowel, a clock, bottles of chemicals," also field guides, maps, notebooks, clothes appropriate for field work—a collection of objects purchased to sample, collect, store and categorize natural specimens, which are taken out of their habitat and confined in boxes and bottles. Most significantly, Hattie finds optical instruments, "magnifying glasses, a microscope, a small telescope," and a camera (76). They were newly developed instruments in the nineteenth century, which enabled a closer, more detailed look at the world penetrating its secrets and promising scientific truth. Not only did they separate and distance the viewer from the objects under scrutiny, they also granted the power to "gaze" and thus to categorize and define the "other" beyond his lens. Photography especially positioned the photographer within the environment. As Susan Sontag noted, "to photograph is to appropriate the thing photographed. It means putting oneself into a certain relation in the world that feels like knowledge—and, therefore, like power" (1973, 4). Edward's quasi-scientific expeditions, conducted with the help of these instruments, reveal his dominating, utilitarian, materialistic approach to the natural environment and to humans. Plants, stones, masks are mere artifacts, interesting for their potential value in dollars. Indigenous people are exotic others, sub-human and inferior to his own civilized ways. When Hattie finds Indigo hiding in their garden, he wants to catch her with a rope, and later he plans to educate her as a maid.

Edward's 'rational' approach to the world, his distance from life, and his desire to control his environment by classification and utilization, has a dark underside. Secretly he is deeply afraid of Natives, women, and his own sexuality. His leg injury, suffered during his failed expedition on the

Pará River, keeps him from consummating his marriage with Hattie; he is afraid of snakes and appalled by the ancient "grotesque" fertility figures in the Italian garden, and tries instead to put all of nature in a box, or a museum, sealing organic life off from growth and decay. Physically and spiritually impotent, Edward's behavior is self-defeating, although it is supported by the then powerful ideology of colonialism of which he is a prominent specimen. It affords him the right to subjugate and kill plants, objects, and "other" human beings, but—as Silko points out—not without a cost to himself. Ironically, his technical instruments, foremost his camera, are not only his tools to fix the world and to keep it distant from himself,[6] they are also his downfall. While taking pictures of orchids in Brazil, he trips over his camera and cannot flee in time to escape from the fire. Later in his life, his own greed and his belief in science make him rely on a quack doctor who first promises the acquisition of rare meteorites in Arizona and then, when he falls ill, the latest medical treatment. His obsessive desire to acquire rare objects and his own blind faith in progress finally kill him.

A very blunt example of the machine's power to destroy an indigenous landscape and culture is the building of the dam in Arizona, witnessed through the eyes of Sister Salt. Deeply influenced by the spiritualism of her tribe, she anthropomorphizes and feminizes the land and the river and laments their violation. "Sister was shocked at the destruction she saw below: the earth was blasted open, the soil moist and red as flesh. [...] The river had been forced from her bed into deep diversion ditches, where her water ran angry red" (211). She finds dead fish and wilted flowers and learns from the Mojave people that the bones of their ancestors buried under the river are exposed and disrespected, inviting witchcraft (212). Native Americans have not only lost their land, but with it a whole culture which depended on an intimate relation with their natural environment. Where white Americans sensitive to natural beauty might deplore the loss of a pastoral landscape, Native Americans must deal with the theft of their homes and their livelihoods. Seen through Native eyes, the machine does not tame a wilderness, but destroys their civilized life, bringing instead disease, poverty and squalor. Among the construction site workers moral values deteriorate; there is gambling, prostitution, violence, and greed. Again the driving force behind the whole enterprise is capitalism: the dam

6 Martin Jay points out that "[t]he camera eye, as monocular as that of the peephole, produced a frozen, disincarnated gaze on a scene completely external to itself" (1995, 346).

is to provide cheap water for Californian agriculture (which Edward would welcome for his desired citrus groves), which, in turn, depletes the soil. And again the destruction of the land goes hand in hand with the extinction of the native population. The technological and scientific achievements of the Modern Era drive the machines that bring progress only to some Americans but are fatal for the continent's first people.

Not all white land use is criticized in *Gardens*. Silko offers the reader a few examples of respectful and spiritual gardening by Europeans, albeit on another continent and resting on a centuries-old, pre-colonialist tradition.[7] The first is Aunt Bronwyn's garden in Bath; the second belongs to *Professoressa* Laura in Lucca. Both gardens as well as parts of Bath show traces of a matriarchal pagan culture which was suppressed and finally defeated by patriarchal powers: Celts, Romans, Normans and others had occupied Bath, each of them bringing destruction and the extinction of the previous culture. Aunt Bronwyn knows the history of the place; she herself is a neopagan and an adherent of animistic nature worship. For example, she keeps white cattle on her property, a Celtic breed, which are symbolic of a female deity with their moon-shaped horns; she also believes in stones that glow and move, some of which have healing powers.[8] The pre-Christian culture in England is thought to have been matriarchal; some stones still bear inscriptions of circles and spirals which Aunt Bronwyn interprets as "the eyes of the original Mother" (265).[9] On the one hand, her old garden thus embodies an almost forgotten female spiritual principle. On the other hand, its variety of plants suggests an ongoing exchange of cultural products among different countries, even continents. Next to indigenous English plants there are "plants from the Americas, Africa, and Asia" (243), all of them carefully cultivated in a foreign climate. The most interesting plant for Indigo is the gladiolus, originally from Africa, which had "undergone a great many changes by the hybridizers" (245). Its history of migration and its adaptability are fascinating for her because they correspond to her traditional ideas about the advantages of exchange that belie notions of purity and exclusivity. It is travel and hybridization which promise development

---

7 There is only one positive example of cultivation in North America, the garden of a Mormon woman which is, however, also destroyed by white soldiers.

8 Joy Porter has shown that Aunt Bronwyn is a follower of Gustav Theodor Fechner, "who venerates 'dancing' stones and sacred groves that house the spirits of the dead" (2007, 65).

9 See Kimberly Roppolo (2007) on the "Sacred Feminine" in *Gardens in the Dunes*.

and growth, for plants as well as for people. Although the metaphor of hybridization could be read as another method of naturalizing Natives, Silko works against such notions by depicting Indigo as a highly 'civilized,' intelligent young girl with a sophisticated knowledge of plant cultivation. She will later set the bulbs into the soil in the dunes, where the flowers will be admired for their beauty, at the same time that some of the bulbs will feed the hungry people. As Laura Coltelli noted, "The concept of hybridization, which pervades the entire novel […] has a primary valence as a quest, as a possibility of interaction among different cultures, and above all as a legacy handed down to us from a past of shared beliefs even if experienced in different cultural contexts" (2007, 187). Grafting, on the other hand, a technique practiced by Edward's father on his citrus plantation in California and by Edward himself, stands in stark contrast to hybridization. It is a method of creating new plants that does not rely on seeds and pollination, but on inserting a cutting into another plant. As such, it symbolizes the violent methods of colonization, or as Deborah A. Miranda aptly observed, "It is not about a process of transformation but amputation and sutures, an artificial and isolated means of change which eliminates most elements of chance and closely controls how, when, and what the tree grows" (2007, 138).

Professoressa Laura's garden in Lucca is also an expression of a once lively spiritual tradition in Europe. She has placed ancient figures into her beds of shrubs and flowers, female fertility Goddesses with naked breasts, minotaurs and centaurs, snakes and frogs, symbolizing, much to Edward's horror, sexuality. To Edward, they belong in a museum, out of the eyes of children, but Laura insists that they are part of a living, organic environment in which they must thrive and decay. She reveres them as artifacts of an animistic, female-based religion which was destroyed by patriarchal culture just like the pagan culture in England and similar to the Native American culture that is being persecuted and wiped out in the United States in the late nineteenth century. The belief in a female principle in Gnosticism was also suppressed by the early Christian church, which Hattie discovered in her theological research for her academic thesis. In the same way that her findings were never accepted by the thesis committee, there has been, according to *Gardens*, an ongoing effort to suppress female spirituality and indigenous knowledge, or what Porter summarized as "aboriginal serpentine matriarchal spirituality" (2007, 193).

The description of the gardens (the dunes, Bath, Lucca) and Hattie's research on Gnosticism during the early years of Christianity emphasize an immanent spirituality that is firmly based in local place but transcends borders at the same time. Indigo learns about parallels in mythology and ways of life that bind disparate cultures together without obliterating their specificity. It was Silko's concern to expose the pernicious effect of national borders. For her, "Regionalism is the hope. [...] Getting rid of all national boundaries. Getting rid of all borders. With regionalism, you do that" (Silko quoted in Arnold, 2000, 186). What she refers to is bioregionalism, in which a geographical area determines the characteristic features of a place where people ideally learn to live on principles of sustainability. That this specific local sense of place does not exclude external forces, but welcomes exchange with other places and cultures is made obvious on various levels in the novel. What is criticized, even condemned, is an attitude of appropriation, of a forceful takeover which exploits and destroys instead of cultivating a hybridized form of new life. Or to quote Isernhagen, "the 'effect of richness' is created in this text through a representation of sites in terms of stratified meanings that programmatically opts for accumulative multiplicity and the transgression of boundaries, rather than exclusionary claims to ownership and the drawing of borders" (2003, 123).

Silko's elaborate descriptions of an indigenous way of life closer to nature illustrate the flaws in the Euro-American pastoral, which rested on the assumption that settlers could create an ideal garden by cultivating the wilderness. From an indigenous point of view, the wilderness never existed because the vast green continent had never been empty but was inhabited by civilized tribes who had their own sophisticated methods of cultivation. She exposes the violence and racism inherent in the visions of the pastoral landscape, which she juxtaposes to supreme forms of cultural relations with the natural environment, based on reciprocity, hybridity, and growth. The machine in her garden is not so much technological progress disturbing an imagined utopian space, it is colonialism and militarism destroying a real culture. Historical facts justify Silko's accusations of the effects of white supremacy. Nevertheless, the question remains whether *Gardens in the Dunes* romanticizes Native ways of life and whether the novel essentializes Native virtues and a female spiritual principle. Silko in fact never criticizes her indigenous characters. They are so firmly grounded in their traditional mentality that they always seem to do and to think what is right. Of course this is partly due to her innocent child narrator, of whom Suzanne Ruta said in an early review of the novel: "Her unshakable assurance gives her the priggish self-assurance of a Victorian child heroine" (1999, 31). It seems that vices belong solely to the white "aliens" and are not part of indigenous life. This prioritizing of Native Americans does not, however, rest on essentialist notions of natural goodness. Their virtues are linked to cultural practices that have been developed in close contact with their natural environment and a careful attention to its material conditions out of a necessity for survival. They have chosen "etiquettes" of behavior and developed norms and rules that assure a civilized, peaceful way of life, which Silko deems necessary for the survival of the planet. It is not reduced to the North American continent, but a part of European culture as well, even if it is an almost forgotten tradition of the past. Just as the Euro-American pastoral implied the possibility of a utopian dream of harmony which had to be achieved in the future, Silko's garden is a lived reality as well as a model for the future. She posits her spiritual, indigenous, inclusionary ways of life as an alternative to the "destroyers." This is not a return to primitivism. Her vision of a more positive reality is multiethnic, open to change, crossing borders, inclusionary, and hybrid, which, taken together, make a strong culture and ideally, a new America.

Works Cited

Arnold, Ellen A. (2000). Listening to the Spirits: An Interview with Leslie Marmon Silko. In Ellen A. Arnold (ed.). *Conversations with Leslie Marmon Silko*, 162–195. Jackson: University of Mississippi Press.

Barad, Karen (2007). *Meeting the Universe Halfway: Quantum Physics and the Entanglement of Matter and Meaning*. Durham: Duke University Press.

Bennett, Jane (2010). *Vibrant Matter: A Political Ecology of Things*. Durham: Duke University Press.

Coltelli, Laura (2007). The Gardens of Memory Between the Old and the New World: "All Who Are Lost Will be Found." In Laura Coltelli (ed.). *Reading Leslie Marmon Silko: Critical Perspectives through Gardens in the Dunes*, 185–205. Pisa: Pisa University Press.

Cronon, William, and Richard White (1986). Indians in the Land. *American Heritage*, 37, 18–25.

Fischer-Hornung, Dorothea (2009). Civilizing Gardens, Fructifying Hybridity, and Cultural Cross-pollination in Leslie Marmon Silko's Gardens in the Dunes. In

Dietmar Schloss (ed.). *Civilizing America: Manners and Civility in American Literature and Culture*, 311–326. Heidelberg: Winter.

Garber, Frederick (1988). Pastoral Spaces. *Texas Studies in Literature and Language*, 30, 431–460.

Isernhagen, Hartwig (2002). Of Desert and Gardens: The Southwest in Literature and Art, 'Native' and 'White'. The Examples of Leslie Silko and Georgia O'Keefe. In Michele Bottalico (ed.). *Literature and the Visual Arts in 20th-Century America*, 173–187. Bari: Palomar Eupalinos.

— (2003). From Identity to Exchange? Some Remarks on the Discursive Construction of Ethno-Cultural Difference and on Leslie Marmon Silko's Gardens in the Dunes. In Helmbrecht Breinig (ed.). *Imaginary (Re-) Locations: Tradition, Modernity, and the Market in Contemporary Native American Literature and Culture*, 107–137. Tübingen: Stauffenburg Verlag.

Jay, Martin (1995). Photo-unrealism: The Contribution of the Camera to the Crisis of Ocularcentrism. In Stephen Melville and Bill Readings (eds.). *Vision and Textuality*, 344–360. Houndmills: MacMillan.

Köhler, Angelika (2002). "Our human nature, our human spirit, wants no boundaries": Leslie Marmon Silko's Gardens in the Dunes and the Concept of Global Fiction. *Amerikastudien*, 47.2, 237–244.

Kolodny, Annette (1975). *The Lay of the Land: Metaphor as Experience and History in American Life and Letters*. Chapel Hill: University of North Carolina Press.

Marx, Leo (2000 [1964]). *The Machine in the Garden: Technology and the Pastoral Ideal in America*. Oxford: Oxford University Press.

— (1986). Pastoralism in America. In Sacvan Bercovitch and Myra Jehlen (eds.). *Ideology and Classic American Literature*, 36–69. Cambridge: Cambridge University Press.

Miranda, Deborah A. (2007). A Gynostemic Revolution: Some Thoughts about Orchids, Gardens in the Dunes, and Indigenous Feminism at Work. In Laura Coltelli (ed.). *Reading Leslie Marmon Silko: Critical Perspectives through Gardens in the Dunes*, 133–148. Pisa: Pisa University Press.

Porter, Joy (2007). History in Gardens in the Dunes. In Laura Coltelli (ed.). *Reading Leslie Marmon Silko: Critical Perspectives through Gardens in the Dunes*, 57–71. Pisa: Pisa University Press.

Ropollo, Kimberly (2007). "We've Got to Get Ourselves Back to the Garden": Indigenous Views of the Life-Death Cycle as Resistance in Leslie Silko's Gardens in the Dunes. In Laura Coltelli (ed.). *Reading Leslie Marmon Silko: Critical Perspectives through Gardens in the Dunes*, 73–89. Pisa: Pisa University Press.

Ruta, Suzanne (1999). Dances with Ghosts. *The New York Times Book Review*, April 18, 31.

Ryan, Terre (2007). The Nineteenth-Century Garden: Imperialism, Subsistence, and Subversion in Leslie Marmon Silko's Gardens in the Dunes. *Studies in American Indian Literatures*, 19.3, 115–132.

Sennett, Richard (2000). Railroads of Arcadia. *Times Literary Supplement* 5094, November 17, 20.

Silko, Leslie Marmon (1999). *Gardens in the Dunes.* New York: Scribner.

Sontag, Susan (1973). *On Photography.* New York: Farrar, Straus and Giroux.

Spicer, Edward H. (1980). American Indians. *Harvard Encyclopedia of American Ethnic Groups.* Ed. Stephan Thernstrom, 58–114. Cambridge: The Belknap Press of Harvard University Press.

Veblen, Thorstein. (2009 [1899]). *The Theory of the Leisure Class: An Economic Study of Institutions.* Oxford: Oxford University Press.

Small Town Love on the Big Screen: The Trope of the Machine in the Garden in Hollywood Romantic Comedies

Klara Stephanie Szlezák

Introduction

"The pastoral ideal has been used to define the meaning of America ever since the age of discovery, and it has not yet lost its hold upon the native imagination" (2000, 3)—thus begins Leo Marx's highly influential, yet not undisputed study *The Machine in the Garden* of 1964. Fifty years after its first publication, the interest in the variations of the pastoral ideal in America is unbroken, and Marx's investigations into the field keep triggering discussions and (re)evaluations. One of the fundamental distinctions which Marx makes early on in his study is the one between two types of pastoralism: "one that is popular and sentimental, the other imaginative and complex" (ibid., 5). While Marx for the larger part of his book elaborates on the second kind, he rightfully calls the first kind "widely diffused in our culture" (ibid., 5). Maintaining a clear divide between 'high' and 'low' culture, Marx asserts that "on the lower plane of our collective fantasy life the power of this sentiment is even more obvious. The mass media cater to a mawkish taste for retreat into the primitive and rural felicity" (ibid., 6). This article traces what Marx termed the "mawkish taste for the retreat into [...] rural felicity" as rendered by one of the mass media, cinema, and more specifically by one of its most "mawkish" genres, the romantic comedy[1]; it suggests, while taking into account the commercial and gendered dimensions of the genre, that the rom-com's take on the pastoral ideal is not sufficiently described in terms of 'mawkishness.' Rather than being entirely devoid of complexity, it oscillates between sentimental and complex pastoralism.

While it is produced globally, the rom-com is predominantly associated with American mainstream cinema made in Hollywood. Some critics ob-

1 In line with the conventions of existing scholarship, this article will abbreviate 'romantic comedy' in the following as 'rom-com.'

serve, and at times consider unjustified, that the rom-com ranges as "the lowest of the low" (Jeffers McDonald 2007, 7), and they provide a number of reasons for the low esteem in which it is held by other critics and which has led to its neglect (Abbott/Jermyn 2009, 1–3). As the rom-com "has typically been perceived as a genre for and about women" (ibid., 7; see also Rubinfeld, xiv) and frequently provides "an excess of feeling which is permitted release in weeping" (Jeffers McDonald 2007, 2), it fits well with Marx's description of the popular kind of the pastoral ideal as sentimental.[2] It thus stands in stark contrast to what Marx would have considered 'art' or 'high culture.'

Yet the gap between 'high' and 'low' culture which Marx upholds and which makes him value certain works more than others has been drawn into question since. Rather than differentiating between separate levels of cultural imagination and production and programmatically categorizing them as either 'high' or 'low' in quality, a more open approach of deeming works originating from the whole spectrum of cultural production worthy of critical examination has taken root. Marx's vision in 1964 of a strict hierarchy among cultural productions moreover reflects a way of thinking that was dominant in academic circles prior to the institutionalization of both cultural studies and film studies as legitimate fields of study. Along the lines of the fading boundaries between 'high' and 'low,' critics rightfully object to dismissing the rom-com too easily. "[The] appeal to audiences of such films," Tamar Jeffers McDonald for instance claims, "is more complex" (ibid., 8) than a simple appeal to a (supposedly female) need for romanticism. Jeffers McDonald does not only point to cases in which "the viewer is inhabiting a position where conflicting pulls of realism and fantasy are operating," but also criticizes that the genre label is too hastily attached, without thorough inquiries into what constitutes the genre and to what extent a given film may show traces of various genres (ibid., 8).

The "more popular kind of pastoralism," as Marx puts it, would merely be a form of escapism if it weren't for "'high' literary culture" (2000, 10). "[In] the culture at large," Marx writes, the pastoral ideal merely instigates "infantile wish-fulfillment dreams, a diffuse nostalgia, and a naïve, anarchic

2 Although rom-coms of course have clear gender implications and although Marx infuses his opposition between the machine and the garden with a gendered dimension (2000, 29), a detailed consideration of this aspect would exceed the scope of this article. For a study of gender roles and relations in rom-coms, see Rubinfeld's *Bound to Bond: Gender, Genre, and the Hollywood Romantic Comedy* (2001).

primitivism" (ibid., 11). When turning his attention to canonical American literary works, however, Marx identifies the "image of the machine's sudden appearance in the landscape" as pervasive and ascribes to it the power to turn "repose" into "tension" and to create "a sense of dislocation, conflict, and anxiety" (ibid., 16). The trope of 'the machine in the garden' generates a complexity that sets the second kind of pastoralism apart from the simplistic, "infantile" kind of pastoralism, which pervades popular culture and "culture at large."

Without a doubt, rom-coms represent popular culture and "culture at large." They are produced to appeal to mass audiences and are widely believed to "provide easy, uncomplicated pleasures" (Jeffers McDonald 2007, 7). The close reading of three late-twentieth-century Hollywood rom-coms—*Baby-Boom*, *Doc Hollywood*, and *Sweet Home Alabama*—shall help illustrate the arguments made in the following: These three films do not only share the theme of a simple "retreat into [...] rural felicity" but they adopt, and present variations of the trope of the machine in the garden. They feature instances of "the machine's sudden appearance in the landscape," which entails tension and conflict. The intrusion and the ensuing conflict are signified and symbolized in these films by the sudden and unexpected encounter between the two protagonists, one urban, one rural, and by the antagonism between them, which of necessity precedes the eventual love relationship. This article explores the specific functions and diverse forms that the trope takes within the case studies. *Baby-Boom*, *Doc Hollywood*, and *Sweet Home Alabama* were released within a period of fifteen years, between 1987 and 2002—a time which "saw a resurgence in the popularity of romantic comedies within mainstream cinema," Claire Mortimer states, "a popularity which shows little sign of abating. [...] The romcom at the end of the twentieth century embraces the traditional in the face of change and crisis in our emotional relationships" (2010, 17–18). As the analysis will show, the three movies to varying degrees use the repertoire of the rom-com as a popular, allegedly particularly 'low-brow,' potentially sentimental, and presumably typically American genre. In the two later movies, the usage of this repertoire reveals an awareness and playful treatment of the genre conventions, which makes possible a reading of these films as self-parodies. All three movies appropriate the trope of the machine in the garden, in very different ways, as a crucial device within their narrative structures, character constellations, and character development, thus testifying to the ongoing meaningfulness and significance for the popular imagination of both the pastoral ideal and the trope of the machine in the garden.

The Rom-Com and the Trope of the Machine in the Garden

Referring to the three films as rom-coms presupposes that there is a definition of the genre according to whose criteria the films can be thus classified. Yet, Tamar Jeffers McDonald underlines that terminological and definitional debates are rare and that "a whole slew of films with very different topics of focus are given the same label" (Jeffers McDonald 2007, 8). In her opinion, the decisive criterion is whether or not a couple and their love relationship are at the center of the film; accordingly, she defines the rom-com as "a film which has as its central narrative motor a quest for love, which portrays this quest in a light-hearted way and almost always to a successful conclusion" (ibid., 9). The problems defining the rom-com largely stem from the fact that both romance and comedy are frequent filmic elements across genre boundaries (see ibid., 8; Mortimer 2010, 3; Krutnik 2006, 132). The rom-com can thus be regarded as a combination of both, with the pursuit of love at its center (Mortimer 2010, 4). Traditionally, the love interest that the rom-com depicts is one of heterosexual love, and the typical 'boy meets girl'-formula (see below) presupposes that the story revolves around a heterosexual couple; yet, while 'mainstream films' still focus on heterosexual relationships, there are also rom-coms with homosexual couples at their center (see Jeffers McDonald 2008, 10; Abbott/Jermyn 2009, 6–7). Further, specific elements help delimit the genre from other, related genres. Jeffers McDonald proposes a useful classification of these generic elements into "visual characteristics, narrative patterns and wider ideology" (2007, 11). Among the visual characteristics, or iconography, she includes settings, props, costumes, and stock characters (such as wedding sceneries, dating outfits, or the initial partner that proves to be the wrong one); she identifies recurring narrative patterns in such small entities as single scenes (involving, for instance, public humiliation to convince the other of one's love) as well as in large patterns spanning the entire movie (like the 'boy meets, loses, regains girl'-pattern); the wider ideology in the case of the rom-com primarily promotes "monogamous coupledom" (ibid., 11–14).

Such definitional considerations help place the three films in question at different positions on a continuum of films that are closer to or farther from a prototypical rom-com. *Baby-Boom* may be furthest removed from a typical rom-com as the actual love quest only begins in the second half of the film, when the protagonist meets her future 'significant other,' and, at least in the first half, is predominantly concerned with the protagonist's professional and domestic problems. In *Doc Hollywood* the love quest is introduced at an earlier point in time but is still paralleled by other themes and narratives. *Sweet Home Alabama*, finally, is most clearly centered on the love quest and relegates everything else to the margins of the story. In all three films, however, one finds typical constituents of the genre, most notably: a man and a woman who eventually fall in love with each other, despite initial obstacles; previous partners who prove unsuitable; and extraordinary events which allow the lovers-to-be the reciprocal approximation which changes their relationship.

Within the realm of Hollywood film, the trope of the machine in the garden is more readily associated with another genre:

> As Leo Marx showed (in *The Machine in the Garden*), American fiction from Nathaniel Hawthorne to F. Scott Fitzgerald concerned itself with the uniquely American experience of a pastoral ideal destroyed by the intrusion of technology. The Western's ambivalent attitude toward progress in general and toward trains in particular situates it firmly within this larger tradition and reveals deep-seated tensions that underlie the Western's account of the West's transformation from a wilderness into a garden. (Belton 2005, 260)

The genre of the Western, as John Belton elaborates, explicitly addresses the connections between the trope of the machine in the garden and the specific geographical, historical, and ideological contexts from which the Western as a movie genre emerged. In the case of the rom-com, these connections are not quite as immediately discernible.

Marx's notion of the pastoral ideal "seeks a balance between the city, which represents corruption, and the harsh and bountiful wilderness" (Umberger). As such, it is demarcated by its divergence from both an (at least seemingly) untouched state of nature and from the characteristics and attributes of urban settings. With respect to film in general, Robert Fish in the introduction to his collection *Cinematic Countrysides* argues that there is a need

> to foreground spatialities of cinema that have tended to be buried beneath existing trajectories of inquiry, for in marking out the relationship between cinema and

space it is the idea of the 'urban', and in particular the 'city', that looms large [...]. Indeed, if cinema functions as both product and instrument of geographical experience, then such experience, it would seem, is of a decisively metropolitan cast. (2007, 2)

Typically, the rom-com unfolds in an urban setting, which often features fancy apartments and chic restaurants (see Jeffers McDonald 2007, 11; Mortimer 2010, 9; Grindon 2011, 20): "The romantic comedy has always been urban and urban-oriented, aggressively, smugly assuming the superiority of city over country" (Brian Henderson quoted in Grindon 2011, 20). In the rom-coms examined here the predominant setting is not urban but rural[3]; and yet, the urban setting plays a crucial role, as a foil, as it were, against which the rural landscape comes into its own. Instead of presenting a differentiated picture, the juxtaposition of the clearly distinct spaces of urban and rural settings reflects an ideological opposition by endowing these spaces with certain mutually exclusive values. The urban settings mainly connote negative aspects that are presented as the downsides of modern city life, such as careerism, opportunism, and superficiality. The rural settings, on the other hand, mainly connote positive aspects, such as integrity, modesty, and sense of community, disregarding, or at least downplaying, economic, demographic, and social problems of small-town America. This opposition prepares the ground for the intrusion of technology into the garden, which then triggers the films' central conflicts.

In all three rom-coms it is the protagonist as a representative of the city who introduces the machine, which is to varying degrees symbolic or allegorical, into the garden, the pastoral setting. To do so, he/she leaves the urban space and enters the rural space; the car ride in particular, taking the protagonist into the garden, is a conspicuous motif found in all three films. With respect to the rom-com in general, Ledger Grindon states: "Trips portray movement and are widely used to imply the transformation of the character in the course of the journey. [...] Because romantic comedy is about the change love brings, these films use transportation settings to portray characters on a moving stage" (2011, 19). In the films, the characters' transformation is closely intertwined not only with a movement from the urban into the rural, but with a transfer of the urban into the rural, which launches the central conflict and thus propels the action, and posi-

3 Other recent rom-coms with a predominantly rural setting are, for instance, *Did you Hear about the Morgans?* (2009), starring Hugh Grant and Sarah Jessica Parker, or also *New in Town* (2009), starring Renée Zellweger.

tions the characters in relation to one another. In the rom-coms under scrutiny here, the trope of the intrusion of the machine into the garden functions both as a means of characterization and as a plot-generating device.

Thus, the urban protagonists and their ideological 'baggage' appear as "sudden, shocking intruder[s] upon the fantasy of idyllic satisfaction" (Marx 2000, 29). Capitalist ideas and agricultural machinery, profit-orientation and luxury automobiles, high-society performances and designer fashion—the cinematic iconography in these rom-coms finds diverse guises for the same pattern. Yet in each of these films, the pattern affects the characters and the plot, and eventually the audience, slightly differently. My analyses of the rom-coms will reveal conspicuous parallels in that they all deviate from the genre's more typical big-city stage and present at the core of their plot a version of the machine-in-the-garden trope. Although all of the films discussed here feature a violent intrusion, which does create tension and conflict, their comedy elements, such as exaggeration, puns, or slapstick, lessen this violence, as does the audience's anticipation of a reconciliatory happy ending. Thus, while the surface of these films is comical, the conflictual encounter between pastoral and urban/technologized elements, personified by the antagonistic but eventually loving couple, deserves a closer look and opens up more profound levels than might be expected from a film genre so universally considered shallow and "formulaic" (Grindon 2011, 1).

Home-Made Capitalism: *Baby Boom* (1987)[4]

Baby-Boom opens with a fast sequence of impressions of the New York cityscape, presenting the protagonist's 'terrain.' The camera then narrows its focus on the city's working women, one of whom is J.C. Wiatt (Diane Keaton), the epitome of a career woman. J.C. has all the 'ingredients' of a successful professional on New York's Wall Street: she has a flourishing career, a lot of ambition, a fancy apartment, and a big office but no scruples, no leisure, no interest in a family, and very limited interest in her

4 According to Mark D. Rubinfeld's list of the "top" 155 Hollywood rom-coms, which ranks films by admission and covers the period from 1970 to 1999, *Baby Boom* ranges as #107, with 6,486,343 admissions, grossing $25,361,603 (2001, 75).

partner Stephen, who is just as wrapped up in his work. Things are going well for J.C. in the corporate world, as her boss offers her a partnership in the advertising firm she works for and assigns her the prestigious Food Chain account. Then, in minute eleven, the unlikely happens: she inherits a distant cousin's baby, Elizabeth, and things go downhill from there. Overtaxed with her new role as a mother, J.C. does not function smoothly as an executive anymore, and when her boss assigns the Food Chain account to a younger, male employee, J.C. quits.[5]

Disillusioned and fed up with the business world, she buys a house in Vermont and sets out for a new life in the country. We see J.C. and Elizabeth ride in a rustic-looking, wood-bodied station wagon from New York City to Vermont. Symbolically, while the New York skyline with its high-rise buildings is fading out, Vermont's rural landscape with its Green Mountains are fading in. Both landscapes overlap for a moment (TC 00:55':11"–00:55':14'), signifying J.C.'s "retreat into rural felicity." On the car ride, J.C. envisions what it will be like in Vermont: there will be a house, an orchard, a pond, a barn, apple pies, and quilts—a whole catalogue of romanticized notions of pastoral life. As Robert Fish notes, "American cinema has a long tradition of coding the rural affirmatively in relation to the values, moralities and landscapes of the city" (2007, 4). *Baby-Boom*, like the other rom-coms examined here, does not break with this tradition. When depicting the intrusion of the urban into the rural as well as the conflict that arises from this, it affirms rural values and moralities,

5 This plot development foregrounds the question of whether women's roles as career women and mothers are compatible in the business world. While J.C. appears clearly overwhelmed by her attempt to maintain her former approach to her job and to do well in her new role as a mother, her male colleagues and her boss appear impatient and little understanding of her challenge to perform both roles at the same time. A key scene is the conversation between J.C. and her boss, just before she quits, when he tells her that he too had to sacrifice part of his family life to be successful in his job, all the while neglecting the fact that unlike him J.C. is a single mother. Her defeat in the face of a corporate system that is shaped according to a male perspective, however, is temporary, and she proves the men who told her that she could not do both wrong at the end of the film. By extension, the fact that she is able to reconcile both roles has to do with her removal to the countryside: her unsuccessful search for a suitable nanny in the city is depicted in great detail, while the audience learns almost in passing that the baby is well taken care of, in J.C.'s absence, by the local nurse, who is also the mayor of the small Vermont town. The countryside seems to present weakened gender hierarchies and a sense of (female) solidarity, which benefits the reconciliation of career and domestic life.

even if the protagonist realizes that her initial preconceptions about rural life do not correspond to her own experiences.

It does not take J.C. long to discover that her expectations were naïve. Milking a cow turns out to be hard work, the house she bought is decrepit, and Hadleyville with its 319, mostly elderly, inhabitants makes her feel lonely. In lieu of a social (and a sex) life, J.C. invents a recipe for baby apple-sauce and spends her snowed-in days filling hundreds of jars. In this phase of the film, the dialogue and action contrast strikingly with the iconography it presents: J.C. claims to be dissatisfied with life in the country; but the 'downsides' of her new life are not only rendered comically, so as to downplay their relevance, but also in a continuously romanticized visual code. The cinematography here includes shots that are likely to appeal to the audience: outside J.C.'s house, there are inches of fresh snow and rows of icicles, glistening in the winter sun; on the interior, there are signifiers of a cozy home, such as a bowl of crisp apples (of which the audience also knows that they are homegrown) and a fire burning in the large open fireplace. It is also now that she first meets the attractive Dr. Cooper (Sam Shepard). The typical initial antagonism between the future lovers (Mortimer 2010, 4) arises when J.C., after having a nervous breakdown, wakes up in a clinic and confides her sorrows to Dr. Cooper only to discover that he is a veterinarian. Feeling betrayed, she pouts and they fight until, after a kiss, a slow dance at the local maple syrup festival, and a night together, they end up as a couple.

The change in J.C. and Dr. Cooper's relationship is connected to J.C. overcoming her frustration with country-life. Paradoxically, she does so by slowly returning to her old life. When J.C., the perennial city slicker, tries to sell some of her home-made apple-sauce at the local store, she runs into other city slickers: a group of rich tourists, for whom rural America is nothing but a "cute"[6] diversion. As they want to buy some of her sauce as gifts, J.C.'s business instincts kick in and she raises the price instantly. What the film demonstrates in the following is a crash course in business strategy. As J.C.'s business picks up, Elizabeth, who was central to J.C.'s decision to turn her back on her New York career, is part of J.C.'s new career as a kind of mascot, giving a face to the brand 'Country Baby.' Accompanied by upbeat music, we see J.C. develop a business plan, do research, introduce a brand name, design a company logo, have labels printed, do

6 In the following, quotations from the films will be given without parenthetical reference.

marketing campaigns and deliveries, have a catalogue distributed, and extend both her production and market. These processes are once more cast in a romanticized visual language: J.C. manages her new business not from a high-rise tower office but on her way to individual country stores, passing by white picket fences and the obligatory white New England steeple, on town squares in personal encounters with her customers, and from within her domestic sphere. We see her in her own kitchen with a few elderly ladies lined-up one next to the other to help her make enough apple sauce to meet the growing demand while the fade-in presents a map of the ever extending market of J.C.'s products (TC 01:18':04"–01:18':13"). What J.C. introduces, against the backdrop of rustic checkered kitchen curtains, is a capitalist division of labor and an assembly line. Despite the romanticized setting inside J.C.'s domestic sphere, this scene, rather than depicting a female small-town community entrepreneurship, implies a hierarchical social structure as well as a distinct profit-orientation. While the women of Hadleyville contribute to the success of J.C.'s business, the film does not clarify whether they do so out of friendship or are paid by J.C. It is furthermore unambiguous that J.C. is the 'boss,' who is head of the business and sets the rules. Furthermore, with the map of New England underlying this sequence, it is made clear that the goal of the women's work is business expansion. The collaborative process depicted here is driven by business optimization and profit-making rather than a sense of community.

J.C. has not retreated to rural felicity after all, it seems, but has established the rules of capitalism in idyllic Hadleyville. A necessary component in this process is technologization. Earlier in the film, J.C. used to pick the apples in her orchard herself (TC 00:57':57"–00:58':10"). Now an aerial view of the apple orchard, or rather apple grove, shows not only the introduction of the capitalist structures of labor division, in the form of professional pickers, but also of technologization, in the form of the machine, literally the machine in the garden (TC 01:20':10"–01:20':15"). When big-city skylines re-enter the iconography of the film, they do so with dollar bills raining down in front of them; we see the skylines of cities across the States overlaid with dollar bills, showing that J.C.'s local enterprise has turned into a nation-wide success story. The film is somewhat ambivalent about this development. On the one hand, this appears as the culmination of J.C.'s efforts and is a take on the potent myth of the self-made wo-/man and the core values of ingenuity, persistence, and hard work. On the other hand, there are hints that this success will collide with J.C.'s new life: we see her change Elizabeth's diapers on her desk while talking on the phone, and a brief conversation with Dr. Cooper about the business and J.C.'s possible move back to New York ends in an awkward search for words.

J.C.'s return to the big-city is only a matter of time, of course. In fact, she never quite left it behind. In the form of the telephone as a symbol and signifier of the continuous link, the big city, its lures and pressures were part of her life all along. A generous offer by the Food Chain brings J.C. back to New York City and to the corporate world, where she feels visibly confident and comfortable. Ironically, it is the outcome of her life far off from New York City—her productivity triggered by her boredom and the business concept that evolved from her rural lifestyle and surroundings—which brings her back: her company is tellingly named 'Country-Baby,' "made in Vermont." However, after the negotiations with the Food Chain, which include the proposition to move the production from Vermont to highly-industrialized Cleveland, J.C. makes a decision against the big money and for her new life: "I'd miss my 62 acres in Vermont…and then there is this veterinarian I met," she says. Strikingly, she answers the Food Chain executives' questions with "yep" and "nope," language associated with rural America. The final shot of the film shows her in her Vermont living room, in a rocking-chair with Elizabeth on her lap, picking a flower from the window sill.

Even if things seem settled by the happy ending of the film, *Baby-Boom* leaves the audience with considerable ambiguities. On the one hand, J.C.'s return to Hadleyville and to Dr. Cooper, her symbolic adoption of the country idiom, and the last view of her in a domestic setting seem to imply that she has turned her back on the big city and all that it entails. On the other hand, there is no indication that her introduction of capitalist mores to rural Vermont will be reversed. At the meeting with the Food Chain in New York City, J.C. had proclaimed: "If you can put Country-Baby on every supermarket shelf in this country, so can I." As pastoral as the iconography of the final shot may appear, the expansion of business looms over the scene. J.C. seems just as much cured of her big-city corporate attitude as idyllic Hadleyville seems tainted by the intrusion of 'the machine' in the form of J.C. and her business practices. The conflict and tension that according to Marx arise from the intrusion of the machine in the garden remain unresolved at the end of the movie. They are instrumental in pushing forward the plot but are silenced in the final shots of the movie in order to provide the happy ending that the genre conventions of the rom-com necessitate.[7]

Of Porsches and Pet Pigs: *Doc Hollywood* (1991)[8]

In just about three minutes, *Doc Hollywood* introduces its protagonist Dr. Ben Stone (Michael J. Fox) as a young, ambitious, somewhat cynical doctor, embodying urban, corporate America. The audience meets him in a busy Washington, D.C. emergency room and learns that he has no friends and is about to leave for Los Angeles to practice plastic surgery. His assertion that he is "not in it for the money" may be read as a foreshadowing of the change he will go through but at that early point in the movie rather reflects his defensive stance in the face of his colleagues' disregard for his new field of practice and is thus little convincing. This introduction is followed by Ben's trip: he is driving from Washington to Los Angeles in his fancy 1956 Porsche Speedster. A notable detail is the camera's zoom-in on his license plate, which reads "A Capital City—Washington, D.C." However, on his journey from one big city to the other, Ben, a reckless driver, in his Porsche symbolically misses the exit that was supposed to take him to the right highway, and toward his lucrative career in Los Angeles, while speeding up to overtake an agricultural vehicle. A truck loaded with hay blocks his view of the sign that announces the exit toward the interstate and Ben drives on without noticing (TC 00:05':47"–00:06':10"). From then on he is lost, erring through the Southern countryside. How far he ventures deep into rural America becomes apparent when he loses radio reception, which distracts him from the road on which, in 'proper' rural fashion, cows led by locals block his way, which leads to the fateful incident. In an attempt not to hit them, he swerves, loses control of his be-

7 Referring to the research of David Shumway, Tamar Jeffers McDonald points to the fact that rom-coms typically end before the newly united couples get married and develop a routine as the depiction of married life and a certain routine would stand in contrast to the romance of the couples' first dates and thus run counter to the rom-com's central ideology of monogamous coupledom, as mentioned above (see 2007, 13). A similar evasion of future plot developments that could harm the initially proposed ideology is at work in *Baby-Boom* with regard to its ideologically laden landscapes.

8 According to Mark D. Rubinfeld's list of the "top" 155 Hollywood rom-coms, Doc Hollywood ranges as #49, with 13,023,938 admissions, grossing $54,830,779 (2001, 72).

loved Porsche, and subsequently crashes it into a white picket fence (TC 00:06':18"–00:07':05"). As his car needs repair and the local 'jurisdiction' sentences him to 32 hours of community service to make up for the damage, the "citified" protagonist (Maslin 1991) is stuck in a small town called Grady[9] and forced to delay his trip to Los Angeles.

As the town urgently needs a doctor, the mayor of Grady tries to convince Ben to stay: "We're still the same sleepy little paradise that big-city shrinks recommend for patients with nervous disorders," he says. And indeed, the film's iconography presents Grady as "an idealization of [...] everything that Beverly Hills isn't" (Rainer 1991).[10] Ben's reluctance to stay and his eagerness to leave are diminished by two factors: he meets (and eventually falls in love with) Lou (Julie Warner); and the local mechanics entirely dismantle his Porsche—Ben's reaction is paradigmatic, given the car's symbolic meaning: "You killed my car. That car was my life."

The symbolic counterpart to Ben's Porsche is Jasmine, Ben's pet pig, given to Ben as payment by a patient who could not pay him in cash. From then on, Ben walks through Grady with his pig on a leash, constantly being complimented on his fine pig. That Ben is still set on going to the big city, despite the evolving relationship with Lou, becomes apparent when he trades the pig for expensive parts that the mechanic needs to repair his Porsche. Yet, it is his interest in Lou that makes him reverse the deal: for the sake of winning Lou over, he gives the pig preference over his Porsche. Throughout the movie, the car and the pig appear as incompatible factors, symbolizing the city and the countryside, respectively.

Lou and Ben's relationship holds no great surprises for the audience who has consumed rom-coms before: it is clear from the beginning that they are meant for each other, that their love story will end well, and that the initial antagonisms are 'part of the game.' What is more interesting is their ambiguous characterization as city and country people, respectively. The audience learns that Ben, despite his "slick city ways," is originally from rural Indiana; Lou, on the other hand, although Ben admits that he "can't picture [her] in the big city," went to college and lived in New York

9 The film gives hints towards Grady's location in South Carolina or Georgia: Ben misses the Interstate in South Carolina, and, later in the film, when he treats a patient, wants to send him to a hospital in Athens. Whether the fictional town of Grady is supposed to be located in Grady County in Georgia is not clarified.

10 In addition, "Grady happens to be a perfectly integrated Southern town" (Maslin 1991) see also footnote 12.

City. The criteria for what makes one a city or a country person are purposely blurred. Maybe it is their ambivalence with respect to these categories that makes Ben and Lou the film's perfect couple. While both have other dating options in Grady, there are no suitable partners for Ben and Lou among the villagers. By implication, these other options do not work out because both Ben and Lou have been altered in a way by their big-city experiences, which set them apart from everyone else in Grady.

Even though the two finally come closer to each other at Grady's Annual Squash Festival, their happy reunion is once more complicated, as Ben is still drawn to the city. He gets into his Porsche, and after a few more complications, leaves for L.A., moves into the obligatory fancy city apartment, and practices plastic surgery, which is presented as tremendously meaningless in comparison to what he did at Grady.[11] As a general practitioner in the countryside his work was a helpful and necessary contribution to the rural community. In Beverly Hills he is one of many plastic surgeons, who spend their days creating artificial appearances. This opposition is significant with respect to the ideologies and values inscribed into the urban and rural landscapes as outlined above. Ben's eventual change of heart makes him leave the big city and the lucrative job to be with Lou in rural Grady. The symbolism and imagery in the final scene are dense. As Ben appears in front of the hospital where Lou works, his pig Jasmine accompanies him. He has noticeably exchanged his suit and tie for casual clothes and sneakers and greets Lou in the words of the locals, "Howdy!" As the three of them walk down Grady's main street, the future of the Porsche remains unknown (TC 01:35':08"–01:36':29").

Like *Baby-Boom*, *Doc Hollywood* is a rom-com whose character constellation and plot progression essentially revolve around the contrast and conflict between the urban and the rural settings. Again, this conflict is brought about by the violent intrusion of the machine, in the form of the Porsche, into a rural small town, characterized by picturesque scenery and likeable, morally integer inhabitants. This intrusion is not only symbolized by the actual machine, the car, but is also personified by the protagonist who represents technologization and the big city. This tension arising from the intrusion translates into the tension between the lovers-to-be, which ultimately must be resolved in the recognition of their mutual love. Yet *Doc Hollywood*'s adoption of the trope is markedly different in many ways, ex-

11 It is clear from the onset of the movie that Ben does not practice plastic surgery as reconstructive plastic surgery but as cosmetic surgery.

emplifying the diversity within the group of 'rural rom-coms.' While the Porsche as a symbol of the disruption is vital to introducing the central (love) conflict and continues to signify tension in the course of the movie, as it is present throughout the film representing Ben's wish to return to the city, the fact that its whereabouts are not mentioned, let alone explained, at the end, suggests a less ambiguous happy ending. The conflict does not potentially persist into the (couple's) future. Ben appears 'cured' of his big-city, big-money ambitions and fully integrated into rural Grady at the end, which more explicitly affirms rural values.

Hillbilly Haute Couture: *Sweet Home Alabama* (2002)

Yet another variant of the conflictual intrusion of the machine into the garden as a trope within the genre of the rom-com can be found in the 2002 film *Sweet Home Alabama*, directed by Andy Tennant. The movie is a prime example of what Claire Mortimer has identified as the characteristic narrative structure of rom-coms:

> boy meets girl, various obstacles prevent them from being together, coincidences and complications ensue, ultimately leading to the couple's realization that they were meant to be together. [...] The narrative often hinges around the central couple, who initially are antagonistic towards each other, but who come to recognize their inescapable compatibility in the face of great adversity and, often, mutual loathing. (2010, 4)

The antagonism between Melanie Smooter/Carmichael (Reese Witherspoon) and Jake Perry (Josh Lucas) is fierce: while they are bound together in a marriage that they entered many years ago in their hometown Pigeon Creek, Alabama, Melanie has moved on (literally, to New York City, and metaphorically, into a new life) and wants to divorce Jake so she can marry Andrew Hennings (Patrick Dempsey), the son of the mayor of New York City. In order to get the divorce, Melanie needs to travel from New York to rural Alabama to confront Jake with the divorce papers and make him sign them as he has ignored her earlier attempts to settle the divorce via mail.

As in the other two case studies, the antagonism between the couple originates from their belonging to and representing different versions of the American experience, urban versus rural. Melanie is a successful fashion designer, whom the audience first encounters in the middle of her preparations for the New York Fashion Week. Her world is introduced with the visual iconography of the metropolis: the New York City skyline, bustling city streets, glamorous scenes by night, a stylish apartment, and a marriage proposal at Tiffany's as the epitome of refinement, luxury, and, potentially, decadence. What follows is the journey that takes the urbanized protagonist into the countryside and thus into 'foreign country'—symptomatically, a character later on says: "People need a passport to come down here." We see a plane land and Melanie drive in her Saab coupé through Greenville, with its stately southern mansions, into Pigeon Creek, with its rather run-down main street. The car ride is accompanied by a switch to country music. Melanie has entered Jake's world, which includes a remote house by the lake, get-togethers in what seems to be the only pub in town, and the obligatory pick-up truck. The antagonism is anticipated before the fighting lovers-to-be even lay eyes on each other.

And yet, the movie from the start subtly weakens this antagonism, which is laid out in the settings and integral to the suspense of the love pursuit, and thus prepares the ground for Melanie and Jake's reunion. Throughout the first part of the movie, Melanie is marked as an outsider in the urban community, despite her flourishing career and her enviable fiancé. At the very beginning, her colleagues comment on her "thick accent"—language serves as a marker of her origins and thus of her 'otherness.' A few minutes further into the movie she is addressed as "Miss Dixie," which intimates that she is not a New Yorker by birth and upbringing.[12] At the same time, as the movie reveals bit by bit, Jake is not an unsophisticated rustic, as the setting may have suggested: he keeps it a secret that he is not only a successful businessman but also, simultaneously, a creative glass designer, the owner of Deep South Glass, designing what one character calls "snooty-falooty glass."

When Melanie returns to her hometown, she is an intruder from the big city who interferes with and disturbs the peacefulness of Pigeon Creek:

12 The contrast between the urban life of New York City and rural Alabama is paralleled by the contrast between North and South. Many passages in the movie refer the audience to lingering antagonisms between the 'Yankee North' and the 'Old South.' The implications of this dimension, however, will not be discussed in detail here as they are not central to the machine-in-the-garden trope that the article focuses on. It is, however, worth noting that despite the movie's allusions to the history of the Civil War an engagement with race is noticeably absent, which is symptomatic of the rom-com's general "discomfort at explicitly confronting race issues" (Jermyn 2009, 24).

she personifies the intrusion into the garden. The film is a little less explicit about this than *Baby-Boom* and *Doc Hollywood*, which at some point have their urbanized characters introduce actual machines into pastoral settings. In *Sweet Home Alabama*, the influence of the big city, its technologization, dynamics, and sophistication, is symbolized by fashion. Fashion here is more than a marker of social status and lifestyle. It works as a signifier of 'technology' in the sense that as a luxury product it encapsulates capitalist logics of production and consumption and thereby creates corresponding hierarchies and dynamics. Not only is the fashion that Melanie wears expressive of her 'new self' as performed within New York City's upper class. It is also the fashion that Melanie produces for a living that makes fashion in the film representative of a decidedly urban, technologized, and consumerist culture standing in stark contrast to the world into which Melanie bursts. Therefore, the characters' costumes are highly meaningful. Melanie's stylishness is a marker of metropolitan lifestyle and sets her apart from the inhabitants of the small Alabama town. This contrast his highlighted in many scenes, for instance when she uses her parents' rusty pick-up truck to drive into 'downtown' Pigeon Creek. Her elegant appearance as she gets out of the truck, wearing high-heeled black leather boots, a tight black dress, designer handbag and shades, does not fit the simple setting of Pigeon Creek's rather run-down main street. Symptomatically, as her old friend Billy Ray drives by in his SUV, he calls out to her: "Wohoo, don't see the likes of you around here these parts. You look like sex on a stick." Later in that same scene, Billy Ray, after hugging her, asks about her scent: "What's that? Calvin Klein? Very nice" (TC 00:32':45"–00:34':45"). Another old friend, herself dressed very plainly, tells Melanie that she looks "like [she] stepped out of a magazine." Yet another scene shows her lying on her bed with a pillow in the background which has the motto "Good clothes open all doors" stitched on it. In a fight with Jake, he accuses her of only caring about "the money, the labels, the shoes."

The references to her fashionable appearance are numerous. But the longer she stays away from New York City, the more her clothes change, signalizing her gradual immersion into the local community. She is still stylish but she now starts to wear cowboy boots, or an all-denim outfit, which allow her to blend in more with the local people. The scene in which Melanie wears the denim outfit shows her at the Pigeon Creek catfish festival,[13] which brings about a change in the love relationship, as the passage ends with a passionate kiss between Melanie and Jake (TC 01:03':33"–01:10':47"). The transformation that has turned her into a big-city girl is slowly reversed while she is staying in the country—"It's not about the money. Not down here," Jake says. The country, Jake seems to suggest, is not about appearances and pretense but about more honest things. Thus, rural America is once more cast in ideological terms as morally superior to and more authentic than urban American.

Citing *Sweet Home Alabama* as another example of the genre's 'love affair' with New York, Deborah Jermyn points out that, despite being set largely outside New York City, the film proves the genre's "compulsion to return to [the New York] landscape" (2009, 11). A closer look at the interrelation in the movie between rural Alabama and the big city reveals that there are in fact two versions of New York: there is the one that Melanie departs from in the beginning of the film and that, above all, is represented by her calculating and manipulative future mother-in-law—after all, and not coincidentally, Andrew's mother is the mayor of New York City. And then there is the one that Melanie and Jake return to. The audience only learns about that other New York in the end titles, which present snapshot-impressions of Melanie and Jake's future: it contains mixed references as some of these 'snapshots' show Melanie in the fashion context, and some depict Jake's Deep South Glass store in New York City; they show the young family on a stroll (presumably) through Central Park, with Melanie dressed conspicuously casually; and they show Melanie, the child, and Jake, wearing a noticeably fancier shirt, in front of his New York City Deep South Glass store, whose sign reads "Deep South Glass, New York, Alabama" (TC 01:40':51"–01:41':48").

On the way towards a happy ending and the couple's realization of their mutual love, the trope of the intrusion of the machine, as symbolized by sophisticated haute-couture, into the garden, the land of lumberjack shirts, is more than a juxtaposition of contrasting settings. It is a significant narrative tool as it makes the central antagonism plainly visible, enhances

---

13 It is a noteworthy detail that all three movies use a town festival as that plot element in rom-coms that suspends the routine and thereby brings about a decisive development in the love relationship, contributing to the dissolution of the conflict. The festival scene as a recurring plot element relates in so far to the urban-rural dichotomy as it underlines the spirit of community presumably only found in small-town communities, where people still know one another and celebrate simple pleasures and occasions, like seasonal, local foods, together.

it, and makes an implicit statement about the couple's future beyond what the movie shows. Melanie's return to Alabama has 'cleansed' her of her big-city attitude, her admitted selfishness. When she returns to New York City with Jake, she takes with it the values of 'the garden' and delights in the simple pleasure that the nuclear family and a walk in the park provide. At the same time, Jake finds a home in New York City and becomes part of the big city. Melanie's intrusion into the garden does not bring about its corruption but instead, through Jake and Melanie's reunion and love, results in a mutual permeation of the machine and the garden in the form of their family from Alabama living in New York City.

Conclusion: Functions of the Machine-in-the-Garden Trope in Hollywood Rom-Coms

"[The] romantic comedy is often perceived to be so obvious in its construction that its components are not analysed," Tamar Jeffers McDonald deplores; "if critical attention is turned on the genre, often what would be legitimated as a trope (a recurrent element) in a genre which has some credibility, is dismissed as a cliché in the romantic comedy" (2007, 10; emphasis in the original). In an attempt to do a bit more justice to a genre that is almost universally shunned by critics, this article proposed a reading of Hollywood rom-coms against the backdrop of Leo Marx's trope of the machine in the garden. *Baby Boom*, *Doc Hollywood*, and *Sweet Home Alabama* testify to and illustrate not only the various "uses of the pastoral ideal in the interpretation of American experience" (Marx 2000, 4); their adoption of the trope of the machine in the garden extends the complexity of both their characters and their narratives.

In such a reading of these films, one can trace the "dialectical presentation of the American experience through the conflicts […] between the pastoral ideal and the proliferation of technology" (Umberger) as investigated and discussed by Marx. Technology, in this context, takes a variety of forms and is inseparable from urban, metropolitan America. Via their urban protagonists, who embody the effects of technologization, the machine is abruptly introduced into the garden metaphorically and its influence makes itself felt through the application of economic and capitalist principles. The effects of technology, and the machine, threaten to corrupt rural morals and ways of life in the shape of capitalism (as exemplified by J.C.'s expansionist business) and consumerism (as exemplified by the Porsche as a luxury car, and by cosmetic surgery and haute couture as luxury products and services), all closely bound to the ideological space of the big city.

Seeing as the rom-com is typically associated with urban settings, the three films chosen for analysis open up a different angle. Cities, and New York City in particular, Deborah Jermyn claims, tend to "constitute the quintessence of faith in the possibility and attainability of romance" (2009, 13). Yet, this claim neglects variants of the rom-com in which the "faith in the possibility and attainability of romance" hinges on a(n) (at least temporary) turning away from the city. When rural instead of urban settings occur in rom-coms, most frequently they are, as Grindon has it,

> accenting the natural landscape of love. [...] What remains constant for the genre, whether it be the metropolitan or the provincial, culture or nature, is the special setting, blessed with humor and in contrast to normal life, which allows repression to fade and emotions free play. (2011, 20)

I have argued that there are films within the genre in which the rural landscape functions in more complex ways that go beyond connotations of "the natural landscape of love." The dialectic between the pastoral ideal and technology, as proposed by Marx, runs through these films like a red thread and turns the city and the country into much more than mere settings: they are central to circumscribing the characters and creating the antagonism between the lovers-to-be that is so vital to the rom-com by imbuing them with antagonistic ideologies. Each location connotes different values and triggers specific associations that correspond to the characters who represent these locations. Therefore the dialectic essentially helps propel the plot as it sustains the seemingly insurmountable differences between two (sets of) characters. As established categories of American culture, the dialectic between rural and urban America evokes such perennial opposites as integrity and corruption, modesty and decadence, inner values and the importance of facades, community and isolation, and thus, ultimately, the presence and the absence of love; the rom-com translates them into popular entertainment. While the trope of the machine in the garden is central to the instigation of the conflict and the relationship of the protagonists, the violence of the intrusion of the machine is diminished by the comedy elements of the rom-coms, and the indispensible happy ending offers an easy settlement of the conflict. It would run counter to the promise of light-hearted entertainment characteristic of the rom-com to leave the audience that it caters to with a sense of unresolved conflict and to deny them a sense of closure.

All three rom-coms share a considerable number of features that establish the dialectic: the lead characters are representatives of the big city; they embark on a journey into rural America; they intrude violently into the pastoral landscape insofar as they bring with them symbols of technology and urban industrial America, which unsettle and disturb the peace of the local community; they are forced by the respective circumstances to stay longer than they want; a central conflict arises between the lead characters and a member of the local community of the opposite sex; yet, as "romantic comedy is founded on the promise that the differences between the man and the woman will ultimately be overcome" (Krutnik 1998, 23), this conflict is certain to dissolve and turn into romance. For that to happen, the conflict between technology and the pastoral landscape, for which the characters stand, needs to be dissolved.

Notably, the dissolution of the characters' conflict, which entails overcoming the dialectic between the machine and the garden, is not effectuated through one prevailing over the other.[14] On the one hand, each plot reveals that the two characters were never as far removed from each other as the audience was at first led to assume: it is in fact the rural character who turns out to have a 'citified' dimension to them due to which they were never full members of the rural communities themselves.[15] This, from the beginning, singles out one character to be the suitable partner and foreshadows the compromise between the opposing ideologies that is necessary to guarantee a happy ending. On the other hand, as the plot advances, the city characters step by step adapt to the 'country ways,' at least to a certain degree, thereby shedding some of their initial urban qual-

14 In the three rom-coms analyzed here the settlement of the conflict is realized either by distracting the audience from a potentially conflictual future with the help of closing images that suggest peace and reconciliation (*Baby-Boom*); or by presenting a character's complete 'conversion' and subsequent adherence to the hitherto rejected ideology (*Doc Hollywood*); or by depicting a final scenario that is essentially a compromise and integrative of previously opposing forces (*Sweet Home Alabama*).

15 In all three films the rural characters are marked as "citified" via education and professional success. In *Baby Boom*, Dr. Cooper, unlike the other villagers, has a university degree, which indicates that he did not spent his entire life in Hadleyville. In *Doc Hollywood*, Lou continues her law studies, which she took up while living in New York City. And in *Sweet Home Alabama*, Jake is revealed at the end to be a creative and successful businessman.

ties.[16] The trope of the machine in the garden serves as a vital means of characterization, character constellation, and plot development in these films. At the same time, it is equally integral to the ways in which the trope functions within this pervasive and popular genre of Hollywood cinema that the dialectic which it establishes is eventually surmounted through the reciprocal approximation of the characters. In the ideal world, in which the rom-com traditionally ends, both rural and urban norms and values complement each other, which is symbolized in the union of the protagonists and legitimized by the love that unites them.

Leo Marx reserved the trope of the machine in the garden and its capacity to produce conflict, tension, and complexity for a part of cultural production that he deemed 'higher' than others. When exploring the adoption of this trope into the genre of the rom-com as a form of popular culture, it becomes apparent that while the rom-com needs to present rather simplified, and thus not always conclusive, solutions to these conflicts in order to allow for its storied to end happily, it does have an awareness of the conflictual potential of the trope and functionalizes it to create genre-specific antagonisms and plot developments.

Works Cited

Abbott, Stacey, and Deborah Jermyn (2009). Introduction—A Lot Like Love: The Romantic Comedy in Contemporary Cinema. In Stacey Abbott and Deborah Jermyn (eds.). *Falling in Love Again: Romantic Comedy in Contemporary Cinema*, 1–8. New York: Tauris.

Baby Boom. (1987). Dir. Charles Shyer. United Artists Pictures.

Belton, John (2005). *American Cinema, American Culture.* 2nd ed. Boston: McGraw-Hill Press.

Doc Hollywood. (1991). Dir. Michael Caton-Jones. Warner Brothers.

Fish, Robert (2007). What are these Cinematic Countrysides? In Robert Fish (ed.). *Cinematic Countrysides*, 1–14. Manchester: Manchester University Press.

16 In all three films this is most manifestly expressed via increasingly casual costumes and via language: in *Baby Boom*, in her last scene in New York City, J.C. uses the countrified 'yep' and 'nope' in a business meeting; in *Doc Hollywood*, Ben adopts the southern 'Howdy y'all' and replaces crude swear words with the more correct substitutes that the villagers ask him to use; the absence and presence of Melanie's southern accent is explicitly addressed in *Sweet Home Alabama.*

Grindon, Leger (2011). *The Hollywood Romantic Comedy: Conventions, History, Controversies*. Malden: Wiley-Blackwell.

Jeffers McDonald, Tamar (2007). *Romantic Comedy: Boy Meets Girl Meets Genre*. New York: Columbia University Press.

Jermyn, Deborah (2009). I Love New York: The Rom-Com's Love Affair with New York City. In Stacey Abbott and Deborah Jermyn (eds.). *Falling in Love Again: Romantic Comedy in Contemporary Cinema*, 9–24. New York: Tauris.

Krutnik, Frank (2006). Conforming Passions? Contemporary Romantic Comedy. In Steve Neale (ed.). *Genre and Contemporary Hollywood*, 130–147. London: BFI.

— (1998). Love Lies: Romantic Fabrication in Contemporary Romantic Comedy. In Peter William Evans and Celestino Deleyto (eds.). *Terms of Endearment: Hollywood Romantic Comedy of the 1980s and 1990s*, 15–36. Edinburgh: Edinburgh University Press.

Marx, Leo (2000 [1964]). *The Machine in the Garden: Technology and the Pastoral Ideal in America*. New York: Oxford University Press.

Maslin, Janet (1987). Baby Boom. Film Review. *The New York Times*, October 07, 1987. 31.07.2014 http://www.nytimes.com/movie/review?res=9B0DE2DA1531F934A35753C1A961948260.

— (1991). A Hollywood Doctor in American Squashland. Film Review. *The New York Times*, August 02, 1991. 31.07.2014 http://www.nytimes.com/movie/review?res=9D0CE0D91F38F931A3575BC0A967958260.

Mortimer, Claire (2010). *Romantic Comedy. Routledge Film Guidebooks*. London: Routledge.

Rainer, Peter (1991). "Doc Hollywood": Southern Exposure. Movie review. *Los Angeles Times*, 02.08. 1991. 31.07.2014 http://articles.latimes.com/1991-08-02/entertainment/ca-83_1_doc-hollywood.

Rubinfeld, Mark D. (2001). *Bound to Bond: Gender, Genre, and the Hollywood Romantic Comedy*. Westport, CT: Praeger.

Sweet Home Alabama. (2002). Dir. Andy Tennant. Touchstone Pictures.

Umberger, Daryl (2014). Marx, Leo. In Miles Orvell (ed.). *Encyclopedia of American Studies*. Baltimore: Johns Hopkins University Press. 03.03.2014 http://eas-ref.press.jhu.edu/view?aid=100.

The Garden in the Machine: Myth and Symbol in the Digital Age

Dietmar Meinel

Traveling through a thick belt of garbage surrounding Earth before passing over a seemingly endless landscape of waste, the opening shots of the computer-animated film *WALL-E* (2008) establish a sublime of environmental destruction. The only moving entity on Earth appears to be a robot, WALL-E, which collects garbage, compresses the junk into cubes, and stacks these cubes into buildings. Through its dutiful work over seven hundred years, WALL-E has rebuilt parts of the skyline of New York out of its compressed garbage cubes and created an eerie semblance of former human order and civilization. This post-apocalyptic scene of global environmental destruction is later followed by a dystopian representation of human life aboard the spaceship *Axiom*. Having withdrawn into space after the ecological collapse of Earth, humankind is portrayed as physically, technologically, and spiritually dependent on machines as people vegetate in this "brave new world" of hyper-modern technology. Although these post-apocalyptic and dystopian visions of the future seem to exclude any notion of a pastoral landscape, the film nonetheless evokes associations with the paradisiacal garden when the mechanical protagonist WALL-E along with its counter-part EVE liberates humanity from its confinement in space to lead a blissful life on Earth. Indeed, the concluding shots of the animated film show the gradual ecological recovery of Earth from the human-made devastation. Throughout its credit sequence, the Pixar Animation Studios film depicts huge spaceships being slowly overgrown by grass and mosses, humans peacefully fishing and sailing in the lush countryside, and WALL-E and EVE harmoniously roaming a decorously pastoral Earth. The depiction of two robots blissfully wandering the paradisiacal garden of Earth, then, seems to echo nineteenth-century popular versions of the assimilation of technological progress into the pastoral and what Leo Marx has described as the myth of the machine in the garden.

In his seminal work *The Machine in the Garden* (1964), Marx famously describes the sudden appearance of the machine in the pastoral landscape as "the root conflict" (2000, 365) in U.S. society. The garden and the machine both function as "master symbols" (ibid., 345) of U.S. culture, "dramatize the great issue of our culture," and represent "the most final of all generalizations about America" (ibid., 353) because of their prominence in nineteenth-century literary culture. While the ending of *WALL-E* can, thus, be situated within the tradition Marx characterizes, the animated film does not only lend itself to a discussion of the Myth and Symbol School. As I will show, the multiple layers of *WALL-E* invite readings from very different perspectives and their corresponding paradigms, readings that highlight distinct aspects of the garden and its relation to the machine. Hence, *WALL-E* can also be read from a revisionist perspective that focuses less on the importance of one unifying "master symbol" and more on issues of imperialism and gender normativity, aspects that Marx did not consider. Similarly, the film's images of global environmental devastation invite a reading of the film that foregrounds the aspect of the transnational, a focus on the ways in which borders are crossed and blurred rather than neatly drawn around a unified, homogenous garden.

On the basis of these three readings and the different aspects they highlight, I will return to Marx's trope of the garden and show how the film also functions as a twenty-first-century update of it: I conclude by situating the battered, grimy, and technologically outdated WALL-E within our contemporary transition from a mechanical into a digital age, a transformation that takes up in many ways the changes and anxieties that Leo Marx traced in nineteenth-century literature and culture. In doing so, I follow Vivian Sobchack for whom the little robot "provides continuity between one stage of our technological development and another, embodying not only the phantasmatic memory of our omnipotent mastery over earlier mechanical technology but also allaying our phantasmatic separation anxiety from it" (385–386). Whereas Marx understood the motif of the machine in nature in nineteenth-century literature as a way to negotiate "an encroaching world of power and complexity, or [...] history" (ibid., 24) in tension with the myth of the pastoral garden, in Sobchack's reading, WALL-E helps to mediate the transition from a stage of humanity's "illusory omnipotence over the world [...] [to] a more mature period in which the human subject recognizes its separation from, yet dependence on, others" (2009, 385). In this reading, however, the myth of the machine in the garden is not only expanded by the twenty-first-century science fiction film but also inverted. Because the animated garden is cinematically produced in the computer—a digital garden in the high-tech machine—the film as such also negotiates a matured human subjectivity. While my exploration of the machine and the garden, thus, speaks to the continued significance of the Marxian trope, the competing readings of the film and the inversion of the machine-in-the-garden-myth also demonstrate its fundamental transformation.

The Robot in the Garden

According to Marx, the cultural, social, and political transformations brought about by industrialization and urbanization in the ninteenth century were negotiated in literature through the contrastive imagery of a "terrain either unspoiled or, if cultivated, rural" (ibid., 9) and a raucous, soiled machine. Although initially debunked as a symbol of despair which represented the increasing alienation of the individual from his or her passions, morals, and social environment (ibid., 176–179), with the advent of the locomotive the machine began to be celebrated in a jubilant, triumphant, and nationalist rhetoric as a potent symbol of economic wealth, political egalitarianism, democratic expansion, and social progress (ibid., 187, 190). After it became a fairly common feature of daily life in the 1830s, Marx goes on to argue, the machine was invested with the faith in "peace, equality, freedom, and happiness" (ibid., 191–192) in popular culture and envisioned as a means "to transform a wilderness into a society of the middle landscape" (ibid., 150) once released from European confinement. Illustrations of the machine melting organically into the countryside portrayed the former as an integral part of the pastoral; in these literary and visual texts "machine technology [...] belong[ed] [...] in the middle landscape" (ibid., 220). The fusion of the (fantasy of the) middle landscape with technology did not merely exalt the corporeal machine to "a transcendent symbol [...] invested with political and metaphysical ideality" (ibid., 206) but also inscribed the physical object with an imperial quality as this American machine was envisioned to "roll across Europe and Asia, liberating the oppressed people of the Old World" and evolve into "a signal, in fact, for the salvation of mankind" (ibid., 206).

As the 'organic' integration of the machine into the pastoral "enabled the nation to continue defining its purpose as the pursuit of rural happiness while devoting itself to productivity, wealth, and power" (ibid., 226), the nationalist and imperialist appropriation of the machine in the garden was, for Marx, "a reactionary or false ideology [...] helping to mask the real problems of an industrial civilization" (ibid., 7). The closing shots of *WALL-E* with the harmonious integration of the robots into an explicitly U.S. American garden seem to echo the triumphant nationalist and imperialist notions of the nineteenth-century myth. And while for Marx the machine had gradually conquered the garden throughout the course of the nineteenth century so much so that "our inherited symbols of order and beauty have been divested of meaning [...] [and] the American hero is either dead or totally alienated from society, alone and powerless" (ibid., 364), in *WALL-E* the American hero does not only liberate humanity from its technological bondage but is personified by the machine.

The first scenes of the animated film show the hero to-be, WALL-E, a Waste Allocation Load Lifter-Earth robot, taking a lunch-box to work, listening to music at work, hanging up its dirty chains when coming home, and turning on the TV after a long day of labor. WALL-E hums and giggles, befriends a cockroach and is ticklish, rocks itself to sleep and is drowsy in the morning. WALL-E longs for love and can be scared—the machine has developed beyond its initial programmed function of waste allocation and appears to be more than a mere mechanical entity. Later, the robot even introduces itself as "WALL-E," signaling the development of an identity over the course of 700 years of compressing and stacking junk.

This identity formation, in addition, is situated within a frontier space. In its arduous process of compressing garbage, WALL-E has rebuilt parts of the New York skyline and thus restored some vague notion of urban order reminiscent of human civilization to the wasteland. This visual depiction of existence on Earth situates the narrative at the intersection of trash wilderness and civilization and therefore localizes WALL-E's identity formation within the specific space of the film's frontier, a hostile space created largely from non-biodegradable garbage.[1] This portrayal echoes the

1 In these first scenes, the remains of civilization are characterized by the little robot in the garbage wilderness. WALL-E represents the last 'living' trace of human life, which has otherwise retreated into space. In the narrative, the polluted planet becomes the frontier where garbage wilderness and human civilization meet as WALL-E turns out to be not a

idea of U.S. identity formation explicated by Frederick Jackson Turner. In his 1893 essay "The Significance of the Frontier in American History," Turner argues that the frontier experience is the vehicle for the "most rapid and effective Americanization" (61) by remaking European immigrants into U.S. Americans through a form of rebirth fostered by the unique qualities of the frontier. For Turner, life on the frontier overwhelmed European immigrants, depriving them of their heritage, and forcing them to adopt the ways of the Natives for survival as the wilderness "strips off the garments of civilization, and arrays him [the colonist] in the hunting shirt and the moccasin" (ibid., 61). The frontier experience peels the colonists "of their past and [gives] them a new and uniform set of American characteristics" such as individualism, self-reliance, and democracy (White 1994, 26). This transformative effect of replacing the pioneers' European garments with those of the Native Americans is reflected in the film. When WALL-E makes its way home, it realizes that its tank tracks are falling apart. The robot then strips itself of the broken parts, and replaces them by appropriating new parts from the debris on Earth—the numerous damaged and salvaged Waste Allocation Load Lifter-Earth robots scattered along the daily path suggest that WALL-E has done this several times over the course of its existence.[2] Through its century-long labor, WALL-E has thus developed a self-reliant identity within the frontier space of the garbage wilderness.

While for Marx the contradictions between the fantasy of the middle landscape and the economic, political, and social experience of the nineteenth century—the tensions between nature and technology—could not be amended due to the increasing anachronism of the machine-in-the-garden trope (ibid., 364), the twenty-first-century animated film reinvigorates this myth. Through its hard, mechanical labor, WALL-E does not only develop an (U.S. American) identity. When the robot transforms Earth into a hospitable planet by cleaning up some of the human-made

marker of the end of biological life but instead the first outpost of humankind about to re-settle Earth.

2 Similarly, WALL-E's home—a salvaged yellow container—also underscores the robot's alignment with early pioneers and the films' frontier theme. Set on a highway overlooking the landscape and separated and above the junk-wilderness, its container-home evokes the tradition of the cabin which "had long been the chief icon of the nineteenth-century frontier, if not of American culture itself" (White 1994, 19).

garbage, when it later liberates humanity from its confinement in space[3] and when it thereby ensures the (re)creation of the pastoral garden at the end of the film, *WALL-E* appears to be evoking a nostalgic longing for a pre-industrial past and an anachronistic yearning for the fantasy of the middle landscape.

From a (Leo) Marxian perspective, then, the "reactionary or false ideology" of the animated film derives from its romanticization of a supposedly simpler, truer, and better life in a pastoral landscape at the end. This escapism—what Marx labeled "sentimental pastoralism" (ibid., 10)—fails to critically explore (or offer practical solutions to) the mounting environmental challenges posed by an expanding (global) consumerism and the (illiberal consequences of an) increasing technologization of everyday life which the film initially presents.[4] With its celebration of the American hero and the resurrection of the pastoral garden, *WALL-E* would be read from a myth-and-symbol school perspective as veiling or clouding the mounting rationalization of human experience by presenting an escapist retreat from the complexity of modern life.

Since the 1970s, however, academics have similarly emphasized the "reactionary or false ideology" embedded in *The Machine in the Garden.* Rob-

3 Moving in their hovering chairs along pre-assigned paths overseen by the spaceship's computer and pampered by robots, the human characters are absolutely detached from their immediate physical environment. Even the communication among humans is mediated through holographic screens that transmit anything from incoming calls to advertisements as all social interaction is technologically mediated. Work and movement are performed by robots which are overseen by the ship's computer. In contrast to WALL-E, humans have lost all individuality and the possibilities of identity formation as they are not able to directly engage with the physical or social world. By overriding the main computer frame of the spaceship, WALL-E initiates the travel sequence which brings humanity back to Earth.

4 This persistence of "sentimental pastoralism" in a pop-cultural text would not come as a surprise to Marx. Only canonical or what he deemed "high literary" writers such as Henry David Thoreau, Herman Melville, and Mark Twain could expose the anachronism of the pastoral myth in the mechanical age and thereby express a critique of the increasing rationalization of human life. Thoreau, according to Marx, feared that "there can be no redemption from a system that makes men the tools of their tools," Melville warned of the "suicidal [qualities] […] to live 'as in a musky meadow' when in truth one is aboard a vessel plunging into darkness," and Twain was even unable "to imagine a satisfactory resolution of the conflict figured by the machine's incursion into the garden" (ibid., 354–355). In contrast to *WALL-E*, these authors foresaw or explored the consequences of an ever-increasing technologization of the social and the consequential demise of the pastoral ideal

ert Sklar, for example, points to the failure of Marx and his intellectual peers to acknowledge "that the analysis of a literary work need not at the same time give an interpretation of the nature of a society and that to account fully for the lives and careers of writers one may first need to understand the social structure in which they lived and worked" (1970, 600). Further, through their concentration on what they considered to be canonical literature, the myth-and-symbol scholars favored one privileged group of writers while perpetuating the marginalization of others (ibid., 599). In exclusively analyzing the writings of a handful of white male authors, Marx and his colleagues failed to recognize cultural diversity, which would have made any assertion of a coherent national culture or literature difficult for them to sustain (Ickstadt 2002, 549). This criticism did not necessarily dispute the challenges posed by an increasing rationalization of human life. Rather, revisionist approaches to American Studies questioned whether canonical authors could actually offer a "radical resistance" against the rationalization of human life as the "[h]ighbrow writers in the tradition of the American Renaissance [...] [were] described as racist, sexist, imperialistic and complicit with the system" (Fluck 2009, 79). In highlighting the classist, gendered, racist, and imperialist ideology in canonical literature, revisionist interventions into American Studies shifted the focus to "marginalized groups and ethnic subcultures" in search of resistance (ibid., 79). The myth-and-symbol approach came under criticism because its scholars postulated a set of dominant tropes, such as the machine in the garden, to be representative of U.S. society. For the following generation of American Studies scholars, however, Marx and his peers failed to see that their assumptions helped to promote what Amy Kaplan and Donald Pease have described as *Cultures of United States Imperialism* (1993). Along these lines, *WALL-E* does not only exemplify the myth-and-symbol approach to culture, but can similarly function to illustrate the ideology critique of revisionist scholars.

Revisionist Interventions into the Animated Garden

For the revisionist American Studies scholars in the 1970s, '80s, and '90s, the inattention to the diversity of identities and to U.S. imperialism in the myth-and-symbol approach represented major instances of its obliviousness to the plurality of U.S. society and a "restrictive and exclusionary" approach to American Studies (Voelz 2010, 20). Even as Marx perceives the representation of the (white, male, heterosexual) American hero in popular culture to be disguising the challenges posed by instrumental rationality, his criticism continues to validate the idea of the (white, male, heterosexual) American hero—whether triumphant, tragic, or disrupted—as the essential figure through which increasing industrialization, urbanization, and commercialization are negotiated. In a revisionist reading of the film, however, the investment in a distinctively U.S. American hero, the multiple references to the frontier, and WALL-E's role as liberator of humanity from totalitarianism pose "the real problem."[5]

With its first shots, the film situates the narrative in the geographical space of the United States. As the camera pans down from space, it homes in on the North American continent and its Eastern seaboard before ending in the topographical region of New York City. As the geographical backdrop of the United States is complemented by the mythical site of the frontier, the setting of the film characterizes the mechanical protagonist as an U.S. American savior of humankind.

5 In the "Afterword" to the re-publication of *The Machine in the Garden* in 2000, Marx responds to this criticism. In tracing the machine-in-the-garden myth in the rebellious rhetoric of the Berkeley student uprising—and in Mario Savio, one of its prominent spokespersons, in particular—Marx attempts to attune his myth-and-symbol approach to the contemporary multicultural political atmosphere (see 383–385). As a consequence, Marx hopes to integrate his initial assumptions about a fundamental dichotomy in US culture without having to revise his theoretical premise. While master myths or symbols continue to be used to negotiate the fundamental transformations in U.S. society, marginalized groups and identities may now find their entry into the limited canon of subversive voices. The updated edition of *The Machine in the Garden* continues to address the "fundamental divide in American culture and society [which][...] separates the popular affirmation of industrial progress disseminated by spokesmen for the dominant economic and political elites, and the disaffected, often adversarial viewpoint of a minority of political radicals, writers, artists, clergymen, and independent intellectuals" (ibid., 383). Consequently, even in his revised version Marx understands American Studies to be a discipline for analyzing a single paradigm (preferably in high-art literature). Revisionist scholars challenged this view with reference to the multiple discourses informing the field. Donald Pease and Robyn Wiegman, for example, write in their introduction to *The Futures of American Studies* (2002), "[t]he contributors to this volume do not assume that the field of American Studies is reducible to the dimensions of a single overarching paradigm, nor do they think that the genealogy of the field can be described as a struggle to predominate among conflicting paradigms" (4).

This humanity similarly exhibits highly monocultural features. Neither in physical appearance nor language or cultural preferences does the film portray the humans living aboard the *Axiom* to be diverse. As the autopilot is programmed to automatically return the spaceship to its home location the *Axiom* descents from space to land near New York, a geographical correlation that further suggests U.S. Americans to function as stand-ins for a global population.

Lastly, the credit scene of *WALL-E* also sustains an imperialist notion when illustrating life on Earth after humanity has returned. In different shots the resettlement of the planet is depicted with humans and robots working closely together in a post-twenty-ninth-century future. Instead of being illustrated in the 'photo-realistic' style of the feature film, these brief episodes or stages of resettlement aesthetically reference various artistic styles of human history starting with ancient Egypt hieroglyphics, the different scenes employ Greek or Roman mosaic, Renaissance charcoal drawings, a multitude of nineteenth- and early-twentieth-century art styles such as Impressionism to end with an 8-bit digital image of WALL-E and EVE holding hands.[6] This journey through art history is, again, deeply rooted in Western traditions as, for example, neither Persian, Chinese, Mayan, nor Japanese cultures are referenced. Although meant to demonstrate the transcultural and trans-species cooperation of all humans and robots, by concluding with the U.S. American savior WALL-E, the film restages notions of *translatio imperii*—the United States represent the logical height of Western culture and civilization which is dated back to Ancient Egyptian, Greek, and Roman empires.

Due to the geographical setting, the mythical site of the frontier, and the monocultural depiction of humans, U.S. culture functions as a synecdoche for humankind in the animated film. Situated within its historical context, *WALL-E* mirrors an imperialist ideology present in the global "War on Terror," in the worldwide network of U.S. American military bases, in the economic dominance of the global financial systems, and in the political hegemony over international institutions such as the United

6 8-bit or 8-bit technology refers to a particular period in the development of computer and video-gaming devices. Popular between 1983 and the early 1990s, the gaming devices utilized 8-bit processors to visualize games in a screen resolution of up to 256 times 240 pixels and a palette of up to 256 colors. These parameters defined the limited scope of the visual imagery used in video games which, by today's standards, would be considered heavily pixelated, blurry, and indistinct.

Nations. For revisionist scholars, however, imperialism is "[n]ot only about foreign diplomacy or international relations," as Amy Kaplan maintains, but "also about consolidating domestic cultures and negotiating intranational relations [..] [since] international relations reciprocally shape a dominant imperial culture at home" (1993, 14).

Whereas the myth-and-symbol perspective would see the appropriation of the machine-in-the-garden myth as evidence for its status as "master symbol" of U.S. culture, revisionist scholars question whether the machine in the garden can actually represent U.S. culture. Or rather, the imperialist features of *WALL-E* demonstrate the ideological dimensions of the machine-in-the-garden myth. For Kaplan, these links between imperialist practices abroad and at home are further evident in their intersections with other normative discourses. Her concept of "manifest domesticity" conjoins the imperialist ideology of Manifest Destiny and Westward expansion with normative notions of gender roles as "[t]he rhetoric of Manifest Destiny and that of domesticity share a vocabulary that runs imperial conquest into spiritual regeneration [...] in visions of geopolitical domination as global harmony" (Kaplan 2002, 31). The concluding shot with WALL-E and EVE blissfully holding hands in the pastoral garden echoes these gendered ideas of spiritual regeneration, geopolitical domination, and global harmony as the names, the visual appearance, and the romance of the two robots establish them as clearly gendered characters.

However, the arrival of a second robot on Earth early in the film invites questions about the representations of gender in *WALL-E* despite its clearly gendered introduction as "EVE." While the bulky, mechanical and technologically dated WALL-E with its visible nuts, bolts, and chains refers to normative conventions of masculine representations, the elegantly floating, hyper-modern EVE with its "alluringly polished, streamlined, even graceful appearance" echoes similar expectations of female physique (i.e. smooth skin) (Montgomery 2011, 10).[7] But by being able to fly and equipped with a powerful weapon repeatedly used to shoot at WALL-E, the female-gendered EVE is technologically far superior and physically threatening to the male cleaning robot. Such a portrayal of an intimidated and vulnerable male figure undermines normative notions of masculinity in these scenes. As WALL-E is immediately charmed and infatuated by the

7 Both robots are also distinguished by their work. While WALL-E is shown to physically engage with the environment by compressing garbage into cubicles and building skyscrapers, EVE simply scans the environment using a ray.

visitor, the male robot attempts to hold hands with EVE and hopes to initiate a romance for which the garbage compressor had been yearning all along (see Bernard 2011, 58). At the same time, WALL-E's passion for a video recording of *Hello, Dolly!* also conjures stereotypical notions of homosexual men enthusiastic about Hollywood musicals (ibid., 54).

WALL-E's investment in the film does not only function as one example of the ambivalent gender roles exhibited by the robots but also signifies "the unnaturalness of romance" as the male robot has acquired its longing for love from a cinematic text: "romance is not inherited; it is learned" (ibid., 54). Consequentially, WALL-E's appropriation of the heteronormative, yet female-connoted cultural practices from *Hello, Dolly!* undermine stereotypical representations of heterosexual male identity just as much as the aggressive behavior of EVE defies stereotypical heterosexual female identity representations in order to highlight romance, gender, and identity as performances.

After the two robots have to take refuge from a sand storm in WALL-E's container-cabin, WALL-E eventually succeeds in calming the aggressive and suspicious EVE. Not only does the male cleaning robot introduce the female drone to the little pleasures of life such as popping protective bubble-wrap, he also teaches her romance by showing the hand-holding clip from *Hello, Dolly!*. Again, this training in "how to act feminine" (ibid., 57) initially demonstrates the performativity of stereotypical gender roles yet it also already indicates the hierarchical structure of male and female as Vivian Sobchack asserts: "It is only under WALL-E's tutelage and care that EVE becomes reflexively adaptive and creative" (2009, 386).

Their brief romance ends abruptly when WALL-E gives EVE a plant he had found among the garbage and debris. Upon storing the plant in her body, EVE immediately goes into hibernation mode, from which the confused WALL-E cannot awake her. Realizing that he does not understand EVE's sophisticated technology, WALL-E decides to re-stage various romantic scenes with her. To savor his romantic feelings, he gondoliers EVE through a river of mud, burns a heart with their insignias into a steel structure, and watches a sunset with her. Just as the technologically dated WALL-E is shown to be more qualified, knowledgeable, and apt than the high-tech scouting drone EVE in finding the only biological form of life, the immobility and stasis of EVE in these scenes of romance further signal the beginning stabilization of stereotypical gender hierarchies because the female EVE is assigned the role of the "mysterious object of desire who motivates the male lead, but who remains essentially passive herself" (Pramaggiore and Wallis 2011, 329). This ascription and stabilization of hierarchical gender identities is not yet completed since WALL-E remains insecure in the dominant position, clumsily stumbling at the end of this scene.

This ambiguity slowly disappears throughout the latter part of the film, as the representation of a dominant masculinity is eventually elevated above a subservient femininity. While WALL-E liberates humanity from its intellectual and physical confinement aboard the *Axiom*, EVE is reduced to the role of an aide. In a pivotal scene, EVE declares WALL-E to be her new prime directive, thus rejecting her prior programming and seemingly asserting her independence, only to be immediately told by WALL-E what to do; she thereby acknowledges a male-centered hierarchy. Lessened to merely learn from and support WALL-E, EVE also skillfully demonstrates that "she has internalized many of the characteristics of the feminine, nurturing the injured WALL-E" (Bernard 2011, 60) after he has been severely damaged. Towards the end of the film, normative gender ascriptions and hierarchies, hence, supplant the portrayal of initially ambiguous and instable identities.

With its stabilization of gender-norms and its imperialist images of a recultivated Earth at the end, *WALL-E* can be read as a highly ideological text. The machine in the garden is, thus, not only updated to a contemporary medium but also exhibits those imperialist and normative qualities against which revisionist scholars have intervened. Instead of locating a subversive potential in 'high literary' works, revisionist scholars shifted their hopes for ideological resistance to "minority groups on the margins or outside of the system, who have not yet been fully submitted to processes of subjection" (Fluck 2009, 74). This continuous search for an outside of and resistance to ideology did not only lead revisionist scholars to investigate the social and cultural margins of U.S. society; in "a logical extension of this very trajectory," transnational scholars subsequently have begun searching for "potential spaces of resistance [...] no longer at the social margins but beyond the nation's physical borders" in recent years (Bieger, Saldívar, Voelz, 2013, ix). This shift towards analytical frames of reference beyond the nation has the potential to break up any homogenous and monocultural notions of a (national) garden with clearly delimited boundaries.

Machines in a Transnational Garden

The vivid images of a world completely annihilated by consumer products in *WALL-E* foster a transnational perspective and invite questions about what Shelley Fisher Fishkin has described as "environmental justice, both at border sites and around the world" (2005, 31). In articulating a programmatic outlook for a transnational American Studies, Fisher Fishkin encourages to interrogate the presumed "'naturalness' of some of the borders, boundaries, and binaries that we may not have questioned very much in the past [...] [and to] probe the ways in which they may have been contingent and constructed" (ibid., 22).[8] One pertinent feature of a transnational research program will be the exploration of "global environment[al] [phenomena] and the ways in which U.S. environmental policies both impact and respond to environmental problems around the world" simply because "pollution knows no borders" (ibid., 31).

With its opening shots of a polluted Earth completely encapsulated by garbage and its camera flight across endless landscapes of devastation, the animated film addresses these transnational challenges posed by environmental destruction. Because this global ecological catastrophe is attributed to the mindless consumption of a highly materialistic and commercialized society, the animated feature exposes the responsibility particularly of the wealthy Western nations for this dystopian future. The fundamental theme of environmental annihilation through unfettered (Western) consumption, thereby, mirrors one vital subject of the transnational turn in American Studies.

In addition, the credit sequence with its portrayal of the resettlement of Earth suggests a transcultural and trans-species collaboration of all humans and robots. Although exclusively Western art movements are referred to in the shots, these images sustain an appeal to forms of cooperation beyond the nation nonetheless. Since national borders do not exist in this post-twenty-ninth-century future, the inherent transnational dimension of the paradisiacal garden may also come to question the present 'naturalness' of

8 Beginning each section of her transnational manifest with the words "[a]s the transnational becomes more central to American studies," Fisher Fishkin lists a variety of transnational interventions into the field. This inventory registers the significance of the "comparative study of race and racism" (ibid., 23), the conceptualization of the United States as a "social, political, linguistic, cultural, and economic crossroads" (ibid., 22), or the demand to explore "borderlands both inside and outside the United States" (ibid., 31) to name but a few.

borders and boundaries. When its opening shots and concluding images are read as a call for transnational cooperation in the face of environmental annihilation, *WALL-E* can function as a site contesting national modes of thinking and seems to portray the United States as merely one "participant in a global flow of people, ideas, texts, and products" (ibid., 24).

When contrasting this transnational perspective with my revisionist reading, however, frictions between the transnational and the imperial moments in the animated film become visible. This tension particularly derives from the nationalist qualities of the machine-in-the-garden myth. The pastoral in *WALL-E*—which transcends the garbage wilderness and the totalitarian *Axiom*—mirrors the exceptionalist tradition of envisioning the United States as "an imaginary land free both of […] oppression and frontier savagery" (Marx 2000, 72). The images of this middle landscape as "a chaste, uncomplicated land of rural virtue" (ibid., 141) endowed with the "moral position" to represent "man's [sic] best hope" (ibid., 101) echo nineteenth-century nationalist ideas in the twenty-first-century text. This semblance suggests the expansion of a distinctively U.S. American symbol to function as a global trope of individuality and self-reliance. If the machine is "not mere scientific or technological development in the narrow sense, but the implacable advance of history" (ibid., 252), history is imagined in *WALL-E* to be the globalization of U.S. exceptionalism.

This imperial quality of the machine in the garden in the film demonstrates the underside of the transnational turn and its incorporation of sites, archives, and phenomena beyond the U.S. nation. Instead of being a liberating site of knowledge, tolerance, hospitality, or resistance, transnationalism can also function

> to enshroud the structural injustices of the contemporary global economic order within the cosmopolitan ethos of a transnational democracy that had not yet materialized […] [as] these highly desirable normative attitudes facilitated the misrecognition of the asymmetrical exchanges of power that in fact took place between transnational and indigenes, migrants, and itinerant laborers at these junctures. (Pease 2011, 15)[9]

9 Not only within the geo-political sphere does the transnational give reasons for caution. Emory Elliott underlines the imperialist qualities of a transnational American Studies if traditional structures of authority and status are maintained within the field. After all, "the place of the United States relative to the rest of the Americas also pertains to the institutional politics of U.S. academic institutions" (Elliott 2007, 9).

Consequentially, a transnational reading of WALL-E may still end up (re)producing U.S. imperial practices on screen, just as a revisionist reading has to acknowledge the ideological instability of the animated film. Nevertheless, the machine in the garden continues to function as a potent symbol: Rather than merely expanding or subverting the nineteenth-century myth, however, the film does not only update the machine and the garden to a twenty-first-century context but, in doing so, transforms these symbols.

The Garden in the Machine

Whereas in the initial encounter of the machine and the garden, the former is described by Marx to be "a sudden, shocking intruder upon a fantasy of idyllic satisfaction" (2000, 29), the animated film establishes an environment explicitly devoid of pastoral nature. From the images of Earth enclosed by trash to the New York skyline made from compressed garbage cubes, the only animate being in this world barren of biological life is the machine. Although WALL-E stumbles upon a variety of human-made twentieth-century consumer objects throughout his journeys, only the little green plant he encounters one day displays the function Marx ascribes to the machine: the little seedling "is made to appear with startling suddenness" and operates as "a sudden, shocking intruder" into the idyllic daily life of the machine (ibid. 15, 29).[10]

As the little seedling designates the habitability of Earth, prompts humanity to escape from its technological imprisonment in space, and represents the romantic connection between WALL-E and EVE, nature symbolizes the (re-)transformation of the machine (-made) world in the animated film. In fact, because of its dutiful work, the garbage compressing robot actually produces the possibility for nature to recover—and thus for the (re)creation of the pastoral garden at the end of the film.

Because of this inversion of technology and nature—the garden manufactured by the machine—*WALL-E* continues to characterize the machine as a savior figure. As a mechanical redeemer from the technologization of

10 Similarly, after humanity has returned to Earth, the camera zooms out to unexpectedly expose fields of green vegetation as nature surprisingly and shockingly intrudes on the devastated landscape.

daily life and its dystopian potential, the robot may refer to what Leo Marx has condemned as a false ideology. While the machine as savior seems to be the logical consequence of the mounting technological interpenetration of human life in the twenty-first century, the visibly mechanic and thus technologically simplistic WALL-E can also "serve to mitigate and soften the terrifying complexity of the technology and the inscrutable amount of bytes, data and circuitry required to render Pixar's animated worlds" (Montgomery 2011, 12). For Coleen Montgomery in "Woody's Roundup and Wall-E's Wunderkammer," the digitally animated world of the film stores and preserves an increasing amount of outdated objects, spaces, and practices for which the garbage-compressing robot with its visibly rusty bolts, corroding screws, stained nuts, and dirty chains is a pertinent signifier. In this sense, the symbols of the machine and the garden are adapted to a twenty-first-century context of hyper-technologization, global environmental destruction, and the vanishing of the mechanical era. As a film programmed on computers, then, *WALL-E* mediates these issues and anxieties by "exploring the capacities of digital media to operate as an archive for those elements of material cultural [sic] rendered obsolete by new technologies [...] [and thus] facilitates identification and association with the digital" (ibid., 12). This normalization of the "terrifying complexity" of digital technology is fostered by a narrative which inverts the machine-in-the-garden myth. The garden is not threatened but manufactured by the machine—in the twofold sense of WALL-E fostering the creation of the garden and the computer (machine) producing the pastoral imagery in *WALL-E*.

Whereas Marx understood the incursion of the machine into nature in nineteenth-century literature as the response of a basically agricultural society to industrialization, the filmic narrative helps to mediate the present transition from a mechanical into a digital era—a transition characterized by what Vivian Sobchack describes as the maturation of human subjectivity:

> Ever present and usually beaten-up and dirty from constant use, the transitional object's value lies in its significant and benign role of allaying separation anxiety—this by providing continuity between a period in which the human subject experiences an illusory omnipotence over the world and others and a more mature period in which the human subject recognizes its separation from, yet dependence on, others, whom it finds it cannot always predict and control. WALL-E, indeed, is just such a transitional object [...]. He thus provides continuity between one stage of our technological development and another, embodying not only the phantas-

> matic memory of our omnipotent mastery over earlier mechanical technology but also allaying our phantasmatic separation anxiety from it, and enabling our shift to an electronic technology that seems, again phantasmatically, often to have a secret mind of its own. (2009, 385–386)

This recognition of our dependency on a phantasmatically independent other is not, however, exclusively mediated through the narrative of *WALL-E*. Although the film aims mostly for a photo-realistic depiction of its characters and mise-en-scène, the animated quality of *WALL-E* never vanishes from sight. Due to its digital aesthetics, then, the film alludes to its production by an "electronic technology that seems to, phantasmatically, often have a secret mind of its own." Not only WALL-E but also *WALL-E* offers continuation in the shift from illusionary omnipotence to human interdependence as the digital imagery of the bucolic garden and the technologically outdated machine both refer to the modern technology essential for their production.

The computer-animated *WALL-E* provides further continuity through its camera work. As a digital text produced completely on computers—thus without cameras and other traditional film equipment—, the out-of-focus shots, frantic zooming, and shaky camera movements *WALL-E* employs appropriate live-action film strategies and a handheld camera aesthetic. The amateur or documentary quality of some of the scenes suggests the familiarity and simplicity of home video filming although the film has been produced with the help of highly sophisticated technology. In this sense, *WALL-E* acquires a transitional quality by linking familiar forms of analog film making to novel forms of digital visual production via sophisticated computer machinery.

This transitional element is highlighted once again in the concluding shots. As the credit scene ends with an 8-bit image of WALL-E and EVE holding hands, the film situates digitally produced imagery within an art history which dates back to ancient Egypt and early forms of human civilization. Again, these shots re-assure a continuation between different periods in human history and an era characterized by digital (computer) technology. Consequentially, since the garden is not merely produced *by* the machine but also *in* the machine, the animated quality of the film functions to mediate the transition from fantasies of human omnipotence (and dominance over the machine) to the recognition of our dependency on it.

In her essay "Animation and Automation," Sobchack, then, does not consider the symbol of the machine as a reactionary myth for a supposedly simpler past, a smokescreen for the global expansion of U.S. imperialism, or a liberating narrative within a transnational context. Fascinated by "the incredible effortfulness of being" as her subtitle states, Sobchack instead offers an anthropocentric reading of *WALL-E* which negotiates human dependency in an increasingly technologized world through the figure of WALL-E; and thanks to its animated quality, *WALL-E* functions as a similar mediation.

Conclusion

In reading *WALL-E* as an extension of the machine-in-the-garden myth in the twenty-first century, I offered competing readings of the animated film using different paradigmatic approaches. While a (Leo) Marxian approach emphasizes that the film's pastoral serves to conceal the mounting rationalization of human life, a revisionist reading foregrounds its (re)production of imperialist and gender-normative ideologies; a transnational perspective, finally, analyzes the implications that the global challenges of environmental destruction pose. On the one hand, the film thus serves to negotiate a number of contemporary concerns, ranging from the increasing technologization of human life to the extensive environmental consequences of Western consumerism. On the other hand, by inverting the relationship of the machine and the garden, the animated film also mediates the current transformation from human (phantasmatic) omnipotence to interdependence to offer a vision of what Sobchack calls a matured human subjectivity which acknowledges its dependence on non-human others.

In one sense, then, *WALL-E* appears to demonstrate the continued significance of the machine-in-the-garden myth Marx analyzed in his seminal book. Since the film portrays the garbage compressing WALL-E as a U.S. American hero and celebrates the recreation of the pastoral garden in the geographical space of the United States, the machine-in-the-garden myth still functions to negotiate issues of national identity, political emancipation, and social progress. Although Marx questioned the validity of such an unequivocal idealization of the machine in the garden, for him only the (white, male, heterosexual) American hero functioned to negotiate the consequences of industrialization, urbanization, and commercialization. With the revisionist interventions into the field of American Studies the ideological dimensions of the myth-and-symbol-school become visible: In *WALL-E* the imperialist and gender-normative dimensions of this perspective are particularly pertinent in the portrayal of humanity as exclusively U.S. American and the hierarchical relationship between the male WALL-E and the female EVE. Because the animated film, however, introduces the American hero to-be as a cleaning robot captivated by Hollywood musicals and longing for romantic love, conventional gender roles appear unstable in the beginning. This instability is furthered by the eco-critical perspective *WALL-E* encourages. As excessive consumption leads to the environmental annihilation of Earth, the animated film offers a transnational perspective warning about the global consequences of Western lifestyles in particular. As each of these competing readings of *WALL-E* captures one vital feature of the animated film, it becomes clear that none of them exhausts the film entirely.

In portraying the technologically outdated robot as a savior of human liberty and a creator of the paradisiacal garden, the animated film also negotiates contemporary anxieties about the mounting technologization of daily life. As a transitional object, Vivian Sobchack asserted, the digitally produced figure of the mechanical, beaten-up, and grimy WALL-E helps to mediate the shift from fantasies of human omnipotence over its machines to experiences of dependency on these nonhuman others. With its appropriation of a live-action camera aesthetic and its situating digital animation within a Western art history genealogy, *WALL-E* functions in a similar fashion. In both instances, however, the robot and the computer machine do not invade a pastoral garden; instead both foster the creation of a (digital) garden.

The contemporalization of the machine in the garden in the animated film validates the continued relevance of the myth as one vital trope of U.S. culture. Yet, its inversion in and by the cinematic text also encourages a reading of the garden in the machine beyond what Leo Marx suggested in his work. Guided by the revisionist interventions and the transnational turn in American Studies, my reading of the garden in the machine signals the continuities and transformations of these pertinent symbols.

Works Cited

Bernard, Carol A. (2011). Performing Gender, performing Romance: Pixar's *WALL-E*. In R. C. Neighbors and Sandy Rankin (eds.). *The Galaxy is Rated G. Essays on Children's Science Fiction Film and Television*, 53–63. Jefferson: McFarland and Company.

Bieger, Laura, Ramón Saldívar, and Johannes Voelz (2013). The Imaginary and its Worlds. An Introduction. In Laura Bieger, Ramón Saldívar, and Johannes Voelz (eds.). *The Imaginary and its Worlds*, vii–xxviii. Hanover: Dartmouth College Press.

Elliott, Emory (2007). Diversity in the United States and Abroad: What Does it Mean When American Studies is Transnational? *American Quarterly*, 59.1, 1–22.

Fisher Fishkin, Shelley (2005). Crossroads of Cultures: The Transnational Turn in American Studies. Presidential Address to the American Studies Association, November 12, 2004. *American Quarterly*, 57.1, 17–57.

Fluck, Winfried (2009). *Romance with America*. Heidelberg: Winter.

Ickstadt, Heinz (2002). American Studies in an Age of Globalization. *American Quarterly*, 54.4, 543–562.

Kaplan, Amy and Donald Pease (eds.). (1993). *Cultures of United States Imperialism*. Durham: Duke University Press.

Kaplan, Amy (2002). *The Anarchy of Empire in the Making of U.S. Culture*. Cambridge: Harvard University Press.

Marx, Leo (2000 [1964]). *The Machine in the Garden. Technology and the Pastoral Ideal in America*. Oxford: Oxford University Press.

Montgomery, Colleen (2011). Woody's Roundup and Wall-E's Wunderkammer. Technophilia and Nostalgia in Pixar Animation. *Animation Studies*, 6, 7–13.

Pease, Donald, and Robyn Wiegman (eds.). (2002). *The Futures of American Studies*. Durham: Duke University Press.

Pramaggiore, Maria and Tom Wallis (2011). *Film. A Critical Introduction*. London: Laurence King Publishing.

Sklar, Robert (1970). American Studies and the Realities of America. *American Quarterly*, 22.2, 597–605.

Sobchack, Vivian (2009). Animation and Automation, or, the Incredible Effortfulness of Being. *Screen*, 50.4, 375–391.

Turner, Frederick Jackson (1993). *History, Frontier, and Section: Three Essays by Frederick Jackson Turner*. Introduced by Martin Ridge. Albuquerque: University of New Mexico Press.

Voelz, Johannes (2010). *Transcendental Resistance. The New Americanists and Emerson's Challenge*. Hanover: Dartmouth College Press.

WALL-E (2008). Dir. Andrew Stanton. Pixar Animation Studios. DVD.

White, Richard (1994). Frederick Jackson Turner and Buffalo Bill. In James R. Grossman (ed.). *The Frontier in American Culture*. An Exhibition at the Newberry

Library, August 26, 1994 to January 7, 1995, 7–66. Berkeley: University of California Press.

The Neoliberal Machine in the Bureaucratic Garden: Pastoral States of Mind in David Foster Wallace's *The Pale King*

James Dorson

"Listen to the silence behind the engines' noise." (Wallace 1989, 373)

For a novel that deals with bureaucracy, David Foster Wallace's posthumously published *The Pale King* (2011) begins in a peculiar way.[1] The short first chapter describes a distinctly pastoral landscape of fields, pastures, a river, horses, sunflowers, and cirrus clouds high above. The passage does not resonate with anything Wallace has written previously, nor does it appear to set the theme or tone for what is to come: a hectic mélange of characters and their troubled trajectories into jobs as IRS employees in Peoria, Illinois. Instead of the "hysterical realism" that James Wood (2000) associated with the headlong style of *Infinite Jest* (1996)—a style which, as he writes, "seems to want to abolish stillness, as if ashamed of silence"—the opening of *The Pale King* slows us down to almost a halt. It seems rather counter-hysterical, grounded in simple observation and contemplation. In contrast to the breakneck sentences of *Infinite Jest*, it is written mostly in a short verbless syntax (e.g., "An alfalfa breeze. Socks' burrs") with a minute attention to detail, which creates a sense of presence over plot, stillness over motion (4). The second person narrative voice which guides the reader through the landscape "to the place beyond the windbreak" enjoins us to stop and pay attention to the surroundings: "Look around you" (3). There are sounds, but they are soothing—so faint that the fact that they

1 Since the novel was unfinished at Wallace's death in 2008, it is impossible to say whether the book would have begun in the same way had he lived to finish it. As the book's editor Michael Pietsch remarks in the "Editor's Note" about the manuscript, "there was no list of scenes, no designated opening or closing point, nothing that could be called a set of directions or instructions for *The Pale King*" (vii). Nevertheless, in a PEN roundtable discussion on the novel, Pietsch motivated his choice of the passage that begins the novel by saying that it was "one of several passages through the book that had been marked at various times in David's writing as a possible opening" (2011, 31 min.). All quotes from this novel will be given with page numbers in brackets.

can be heard at all just deepens the overall sense of calm: "Electric sounds of insects at their business", "Dry scratching inside a culvert", "The shush of the interstate off past the windbreak" (3, 4). The clamor of the modern world, its intrusive technology and ceaseless administration, seems remote. The roar of the interstate is reduced to an onomatopoeic shush by the windbreak, sheltering us not only from the wind but from the machine noises of the outside world.

Such a strikingly pastoral opening to a novel about something as apparently non-pastoral as the Internal Revenue Service, and by a writer known not for his use of pastoral conventions but for his struggle with postmodern ones, begs the question, why *The Pale King* begins on such a pastoral note, or, more precisely, what function the pastoral idyll serves in a novel largely about bureaucracy? After all, pastoralism and bureaucracy are usually represented as mutually exclusive. Pastoral connotes ease and leisure; bureaucracy is associated with rule-bound experiences. While a pastoral way of life today is viewed as solitary and non-conformist, a liminal existence on the margins of society, life in a bureaucratic society is viewed as infernally complex and convoluted, mass-produced and mass-consumed, by and for the masses. When Ishmael at the beginning of *Moby-Dick* (1851) talks about the attractions of going to sea, which Leo Marx in *The Machine in the Garden* (1964) saw as a profound expression of pastoral longing (282), it is not to get away from work as such, but to escape the specific drudgery of deskwork: "days pent up in lath and plaster—tied to counters, nailed to benches, clinched to desks" (Melville, 22). This also suggests that pastoralism and bureaucracy are not only mutually exclusive, but that pastoral yearning in the context of modernity is constituted by this very antagonistic relationship with its bureaucratic other. Pastoral does not simply connote an alternative to bureaucratic society but has long functioned as a privileged site for antimodern critique.

In the following, I will argue that Wallace in *The Pale King* reverses the traditional roles played by bureaucracy and the pastoral longing away from it. This at once suggests that these characteristically modern roles have become defunct in post-Fordist society, and that Wallace in his last major work tried to offer us a version of pastoral more appropriate for dealing with contemporary life. In order to examine the function of pastoral idyll in *The Pale King*, I will read Leo Marx and Wallace together with the aim to show both the continued relevance of Marx's seminal analysis of pastoral ideology as well as Wallace's continued reliance on a form of cultural criticism that he at first appears to renounce. My argument, in short, will be that *The Pale King* pastoralizes bureaucracy at a time when the modern bureaucratic outlook that shaped much of the twentieth century is being undermined by the neoliberal marketization of society, but, moreover, that this pastoralization of bureaucracy, instead of preserving its underlying organizational tenets, rather takes a form that is congruent with the very managerial practices that have smoothed the transition from progressive to market values.

Pastoral Mindscapes

Does it still make sense to use the concept of the "pastoral" other than to signify a form of escapism in a world that has been defined as being "after nature"?[2] This is the first question that inevitably arises when considering the function of pastoralism in *The Pale King*. Leo Marx defined the pastoral impulse as "a desire, in the face of the growing power and complexity of organized society, to disengage from the dominant culture and to seek out the basis for a simpler, more satisfying mode of life in a realm 'closer', as we say, to nature" (1986, 54). But what happens to this impulse in the fully administered world of late capitalism in which nature no longer poses a limit to organized society, where it is no longer possible to simply "disengage"? Anticipating this problem in *The Machine in the Garden*, Marx had called for "new symbols of possibility" more adequate for a nuclear age (1964/1967, 365). His study may as well have been titled "From the Garden to the Machine" for it delineates a trajectory from the first literary appearance of a "machine" in the "garden"—"the disturbing shriek of a locomotive" (ibid., 16) noted by Hawthorne in 1844—to the final transformation of nature by the industrialization process into a "garden of ashes" in *The Great Gatsby* (1925). Because Marx had defined the pastoral garden as "a symbolic middle landscape created by mediation between art and nature" (ibid., 71), the destruction of the polarity—wilderness on the one side, civilization on the other—that the pastoral balance rested upon meant that both wilderness and garden had now been subsumed by the machine side of the equation. With the triumph of mechanized society, the

2 See Frederic Jameson's definition of postmodernism as "what you have when the modernization process is complete and nature is gone for good" (ix).

pastoral ideal for Marx could no longer function as an alternative political imaginary. If it had once served as a symbol of possibility in the "New World" due to its vast reserves of pastoralizable wilderness, the unrelenting industrialization of the U.S. had turned the pastoral garden into a symbol of impossibility.

Like his teacher Henry Nash Smith's discipline-founding study *Virgin Land* (1950), *The Machine in the Garden* was Marx's attempt to bury the "ghosts of outmoded idealisms" (Smith, 195). For critics associated with the Myth and Symbol School such as Smith and Marx, who had come of age intellectually during the political struggles of the 1930s, "outmoded idealisms" meant an obstacle to progressive reform and as such were a conservative force in society. Marx's focus was on literature that exemplified a "complex pastoralism" that acknowledged the incongruence between "garden" and "machine", and which thus countered a "sentimental pastoralism" in popular culture and national mythology that disregarded the incursions of social reality. In retrospect, however, Marx claimed that it was not long after the publication of *The Machine in the Garden* that he realized how he "had underestimated the depth and durability of the conflict I wrote about" (1999/2000, 493). While he had viewed pastoral ideology as a form of false consciousness serving the political Right, the growing momentum of the counterculture in the immediate aftermath of the book's publication demonstrated for Marx how pastoral ideology had been revived with a vengeance on the Left. If the appearance of the machine in complex pastoralism had served as a counterforce to naïve pastoral beliefs, now the pastoral ideal of social detachment had been mobilized by the new social movements of the 1960s and 1970s as a counterforce in its own right to what was considered the repressive power of the "state apparatus".

Yet the resilience of the pastoral ideal as a site of cultural longing in the face of altered social conditions was implied by the very generality of the terms with which Marx defined pastoralism. From the outset in *The Machine in the Garden* he had made clear that the pastoral garden was not contingent upon any actual landscape, but was a "landscape of the psyche" (28). On Hawthorne's notes from Sleepy Hollow, Marx writes that "[h]e is describing a state of being in which there is no tension either within the self or between the self and its environment" (ibid., 13). For Hawthorne, it was "not the matter so much as the feeling behind it" which counted (ibid., 12). His decision to begin with Hawthorne's Sleepy Hollow notes was then not merely because it "prefigures the emergence, after 1844, of a new, distinctively American, post-romantic, industrial version of the pastoral design", but also because it exemplifies that the pastoral design is not about a specific type of location but rather about a specific frame of mind (ibid., 32). "To sit down in a solitary place", Hawthorne begins his notebook entry, but then immediately adds in parenthesis, "or a busy and bustling one, if you please" (1932, 102). The object of attention is not important; what counts is the attention itself. His peaceful mindset is disturbed twice in the short passage, not only by the whistle of the locomotive but also by the "intense hum of a mosquito" (ibid., 104). Both times the incident causes a loss of presence: in the case of the mosquito because he reacts instinctively ("we crush him at once"), and in the case of the train because it transports him to someplace else ("It tells a story of busy men, citizens, from the hot street") (ibid., 104). The incidents pose a problem not so much for what they are—dangerous wilderness and stressful civilization—but for what they do, which is to disturb his sense of presence, his attempt "to observe such noticeable points as the eyes fall upon around you" (ibid., 102). The pastoral garden is thus a state of mind that may be cultivated by paying attention to our immediate surroundings; in other words, by being present in the "now", as the insistent use of the present tense ("Now we look around us") and the injunction to observe ("Observe the pathway") in the notes make clear (ibid., 103).

Conceiving the pastoral ideal in psychological rather than spatial terms meant that in theory it could be achieved in any setting or location provided that it was possible to sustain attention over a protracted period of time. This sense of the pastoral as a mindscape rather than landscape was not only based on the material Marx used but had its roots in the postwar ideological climate in which he wrote. In an essay titled "Notes on the Culture of the New Capitalism" from 1959, Marx had joined the critical chorus of his times in advocating a turn away from a traditional Marxist economic focus on production, since "today the problem of productivity in America apparently is solved", and toward a "socialist humanism" that instead sought to address the psychological consequences of the productive system (112, 113). Industrialization may have brought affluence, but the machine process also brought a "mechanical quality of mind" (ibid., 111). Just as the garden was a "landscape of the psyche", the machine signified a mental state, less a matter of actual technology than of the instrumental rationality that accompanied it. As such, for Marx the most pressing problem facing his generation was not material, but "what our society is doing to our minds", the greatest danger not economic but "that familiar Orwellian nightmare—the vision of a bleak, mindless, undifferentiated culture of robotized consumers" (ibid., 112, 114).

Such evident sympathies for an alienation critique of modern capitalism would seem to have predisposed him favorably toward the new social movements, which took up precisely this strain of criticism, if it had not been for one crucial difference. For Marx the mental pastoral retreat from the mechanized reality of capitalist society did not represent a permanent solution to the problem of alienation. Rather, as *The Machine in the Garden* made clear, the pastoral ideal was a "transitory state of mind" and should not be confused with a viable political alternative (280). In this respect, while his diagnosis of alienation and instrumental reason as the chief threats to modern society coincided with that of the new social movements, he did not share their remedy. If his original definition of pastoralism as the "withdrawal from society into an idealized landscape" may easily have encompassed a radical political ideology, his progressive narrative from "garden" to "machine" effectively confined the pastoral ideal to the imaginary realm of aesthetics (ibid., 10). Pastoralism had begun as a literary mode with Virgil's *Eclogues*, and it was only with Jefferson's *Notes on Virginia* (1785) that "the pastoral ideal had been 'removed' from the literary mode to which it traditionally had belonged and applied to reality" (ibid., 73). Since the Jeffersonian ideal had never been realized, it was Marx's aim to return the pastoral ideal to where he felt it belonged: in the literary imagination. This was for Marx the great merit of the American literary tradition of complex pastoralism, because while it maintained that "the image of green fields is meaningful" as a repository of antimodern longing, it recognized that this was only the case "so far as it is joined to its opposite. In itself the image represents neither a universal condition nor a set of values which can be embodied in social institutions" (ibid., 313).

As such, *The Machine in the Garden* may be read as a tacit response to one of the great precursors of the cultural politics emerging in the 1960s. Published in 1955, Herbert Marcuse's *Eros and Civilization* had been intent on challenging the separation of politics and aesthetics—"the distortion of the aesthetic attitude into the unreal atmosphere of the museum or of Bohemia"—and made the provocative case that traditional institutions should be transformed on the basis of aesthetic values (145). For Marcuse, it was precisely because aesthetic experience was the only realm not yet dominated by the "total mobilization" of postwar society that he envisioned "the liberating force of the aesthetic function [...] as containing the possibility of a new reality principle" (ibid., 70, 148). Like Marx, Marcuse equated the metaphor of the garden with an aesthetic attitude, but unlike Marx, he relocated the vision of a garden that sees "man and nature as joined in a non-repressive and still functioning order" from the irretrievable past to the realizable future, as a model for "the rational reorganization of a huge industrial apparatus" (ibid., 173).

Marx did not articulate his difference with Marcuse in *The Machine in the Garden*, although he did mention his book in passing. In later essays, however, he became more explicit about his critique of cultural politics. Concluding his own revisionist account in "Pastoralism in America," Marx asked whether the reemergence of the pastoral ideal in the new social movements might actually "provide the basis for an effective political ideology" (1986, 66). His answer was that for pastoral ideology to be politically effective, it had to be joined with a more traditional base for class struggle, "the hitherto disadvantaged carriers of emergent values" (ibid., 66). Pastoralism may have been "well suited to the ideological needs of a large, educated, relatively affluent, mobile, yet morally and spiritually troubled segment of the white middle class" (ibid., 40). But if Marx had identified with the alienation critique of the 1950s, his progressive schooling during the 1930s finally determined his view of social change that cultural disaffection alone neither had the ideological coherence nor the organizational will to effect lasting political change. As such, he continued to stand by the words with which he had concluded *The Machine in the Garden*, namely that "[t]he machine's sudden entrance into the garden presents a problem that ultimately belongs not to art but to politics" (365).

We might say that Marx has consistently resisted the literalization of the pastoral ideal on three fronts: its conservative reification in national mythology as "Nature's Nation"; its radical literalization as a model for an alternative political reality; and most recently its ecocentric literalization as a material space independent of its cultural signification, an argument put forth by Lawrence Buell, who has lamented Marx's concept of nature "as a politics of culture rather than as a domain of engagement in itself" (1995, 47). But insisting that the pastoral ideal is a state of mind and not of affairs has allowed the concept to maintain a critical edge as an analytical category. The emergence since the late 1970s of what Luc Boltanski and Eve Chiapello (1999) have termed "the new spirit of capitalism" has shown that the quest for pastoral fulfillment, instead of bringing about a more harmonious reality, has not only been accommodated in the New Economy but has served as the cultural engine for its neoliberal intensification by rejecting progressive institutional frameworks which had held it in check. For Marx, the problem with the cultural politics of pastoralism as a political ideology was that it was precisely that: cultural. Being finally a question of inner rather than outer reality, the pastoral ideal would in principle be feasible under any social conditions. If the problem could be defined as a matter of attitude, so could the solution: as a cultural ideal it did not have to replace the political or economic system but could exist alongside and even justify it. This was the contradiction that early Americanist critics such as Marx had explored by examining how pastoral ideology could coexist with a political system geared toward technological progress and capitalist growth. And this same question may well be asked again today in respect to how various cultural quests for self-fulfillment against the limitations imposed by bureaucratic society have engendered new and more subtle processes of rationalization. Instead of freeing us from an instrumental mentality, the cultural success of the anti-bureaucratic tendencies that Marx identified with pastoralism have helped, in the words of Boltanski and Chiapello, to "facilitate an instrumentalization of human beings in their most specifically human dimensions" (1999, 98).

Pastoralism in the Office

If *The Pale King* begins by establishing a sense of calm by focusing the reader's attention on the surrounding landscape, this pastoral mood is soon shattered. In the next chapter, we are introduced to Claude Sylvanshine aboard a plane from Midway to the IRS center in Peoria, where he works. While the opening chapter sheltered us from the noise of machines, this chapter unfolds entirely in the belly of one: "a terrifying thirty-seater whose pilot had pimples at the back of his neck and reached back to pull a dingy fabric curtain over the cockpit and the beverage service consisted of a staggering girl underhanding you nuts while you chugged a Pepsi" (5). The expansive calm of the first chapter has been replaced by what sounds most like postmodern hyperrealism. Midway is not a middle landscape but an airport in Chicago, and the pastoral suggestion of Sylvanshine's name is tragically ironic. Mirrored in the frenetic style of the prose, the packed sentences and impossible digressions, he is unable to relax or focus "nothing would hold still in his head in all the confined noise" (6). We soon learn that concentration is a general problem for Sylvanshine. For over three years he has been studying to pass the Certified Public Accountant exam, but "studying any one thing would set off a storm in his head about all the other things he hadn't studied and felt he was still weak on, making it almost impossible to concentrate, causing him to fall ever further behind" (9). Since he knows that "[t]he entire ball game, in terms of both the exam and life, was what you gave attention to vs. what you willed yourself to not", he does what he can to manage his attention (12). Using an "anti-stress technique called Thought Stopping", he tries "to pay close attention to his surroundings as a way to avert thought and anxiety" (15, 21). "Another effective concentration device", we are told, "was to summon into one's mind a soothing and low-pressure outdoor scene, either imagined or from memory, which was even more effective if the scene comprised or included a pond lake brook or stream, as water had been proven to have a calming and centering effect on the involuntary nervous system" (12). Yet nothing that he does is able to quiet the "storm in his head", and the chapter ends with him having a panic attack, a feeling of "Total Terror and Disqualification" (24).

Sylvanshine may attempt to invoke pastoral imagery as a becalming fantasy, but the world of *The Pale King* is not conducive to such reveries. Besides the first chapter, nature does not figure as a site of pastoral escape. Almost the entire novel unfolds in the enclosed spaces of planes, cars, elevators, offices, classrooms, and a mental institution. "I learned that the world of men as it exists today is a bureaucracy" (437), one character says, and we are given no reason to think otherwise. The penetration of the rural Midwest by instrumental reason is perhaps best encapsulated by a recurring sign that reads "IT'S SPRING, THINK FARM SAFETY": even spring does not evoke pastoral hopes but only industrial anxiety. While the novel does parody the circular logic and effects of a bureaucratic system—as when such a "farm safety" sign is torn loose during a tornado and nearly decapitates a farmer—and in this way clearly echoes the devastating critique of bureaucratic rationality in a novel such as Joseph Heller's *Catch-22* (1955), the problems faced by the characters in *The Pale King* have less to do with the exasperating recursivity of the system than with the recursivity in their own heads (275 ff.). And unlike Yossarian at the end of *Catch-22*, this is something that they cannot simply run away from.

Talking to one interviewer about modern corporate culture, Wallace has said that "this was our world and reality, the same way the Romantics' world was trees and babbling brooks and mountains and blue skies" (2004, 130). No escape into nature is possible in Wallace's work. Yet at the same time he has made clear that he was not interested in creating "an ingenious mimesis of a bad world", but that the aim of good fiction should be "both to depict this dark world and to illuminate the possibilities for being alive and human in it" (1993, 26). Noting this stress on being "human", which recurs throughout Wallace's work, Paul Giles (2007) has argued that "Wallace starts with abstraction and then uses the human element to subvert rigid technocratic patterns" (333). Giles thus situates his work within "a familiar strain of American pastoral" (ibid., 331). If Wallace does not retreat to Walden pond, his rendition of contemporary life for Giles is nevertheless "one that opens up crevices in the monolithic structures of corporate America and confronts the question of personal authenticity even within a global framework of displacement and irony" (ibid., 341). As such, we might then say that if Marx understood the pastoral impulse as a centrifugal movement away from society, in Wallace the pastoral impulse is defined by a centripetal movement toward the heart of contemporary experience. This search for a still center—instead of a still outside—that may help us weather the storm is displayed both in Wallace's style, which he himself described as "tornadic", and in the narrative arc of *The Pale King*, where characters seek refuge in the IRS—the very epicenter of U.S. bureaucracy—from the clamor in their heads.[3] Perhaps the most poignant metaphor for this endeavor is found in Wallace's early story "Westward the Course of Empire Takes Its Way" (1989/2011). In describing a wheel "spinning in place", the sheer velocity of which makes it look as though it stood still, the story bids the reader: "Hold rapt for that impossible delay, that best interruption: that moment in all radial time when something unseen inside the blur of spokes seems to sputter, catch, and spin against the spin, inside" (ibid., 373).

The search for a still center—"that best interruption"—within the squall of contemporary life is thus not new to Wallace; what is new in *The Pale King* is the role that bureaucracy plays in achieving it. The longest

3 In the foreword to *The Pale King*, Pietsch writes that Wallace in his notes "referred to the novel as 'tornadic' or having a 'tornado feeling'—suggesting pieces of story coming at the reader in a high-speed swirl" (viii). Yet a "tornadic" effect also suggests the possibility that readers may suddenly find themselves in the quiet eye of the storm.

chapter in the novel, and what functions as its thematic linchpin, is the story told by Chris Fogle of how he discovered his passion for the IRS. In contrast to the narrative of progressive estrangement and escape from an oppressive system in *Catch-22*, Fogle's story reverses the typical alienation plot by escaping toward bureaucracy instead of away from it. Beginning in the late 1970s after the counterculture has run its course, Fogle explains how the rejection of the values and norms of the preceding generation—embodied by his father who was "like a robot or slave to conformity"—had led to a new era of conformity in which everyone "appeared nihilistically wastoid and hip" (167, 163). Like his fellow students at college, Fogle describes his former self as "a total conformist to the late-seventies standards of so-called nonconformity" (164). His conversion experience comes when he accidentally finds himself in an accounting class on Advanced Tax. Here he is literally "called to account" by the Jesuit teacher, who holds forth on the values of accounting (233). Echoing a passage from *Infinite Jest* in which Hal Incandenza discusses the bureaucrat as "the hero of non-action" (142), the Jesuit teacher similarly suggests that accountants are modern-day heroes in their ability to face the boredom of routine and in their capacity of sorting out and ordering society's "unending torrent of financial data" (233). As the turning point in the narrative, the talk inspires Fogle "to renounce nihilism and make a meaningful, real-world choice"—in other words, to join the ranks of the IRS (237).

Like the pastoral ideal, accounting strives for balance. The accountant creates order and harmony out of a wilderness no longer associated with nature but with society. The IRS in *The Pale King* not only figures as a safe haven from the data overload of information society but as the means by which to recreate a pastoral balance which has been lost, ironically, precisely as a result of the pastoral longing away from bureaucratized life in the first place. If the counterculture had aimed to recreate society in the image of a garden, *The Pale King* suggests that the outcome was a social and psychological wilderness. Moreover, the very elements that the counterculture had protested—excessive discipline and dehumanizing work—are reinscribed in *The Pale King* as the very means to the pastoral state of mind that they were initially seen to disturb. *Infinite Jest* had already explored the possibility of transcendence through the discipline of competitive tennis: "how to avoid thinking [. . .] by practicing and until everything runs on autopilot and talent's unconscious exercise becomes a way to escape yourself" (173). Evoking images of Taylorized factory work, tennis is described as a form of automation that develops a "machine-language" in players through "sheer mindless repeated motions" (118). *The Pale King* takes this one step further by figuring the routines of bureaucratic deskwork itself as a form of "Self-Storage"—the name of the parkway on which the IRS center in the novel is located—where the deep concentration that characters like Sylvanshine so desperately seek is made possible through rote work that demands your full attention.

This is not to say that Wallace was not deeply ambivalent about ways of blocking thought. In true modernist fashion, the narrative complexity of his own work was meant to counter the regression of art into entertainment. In his 1993 interview with Larry McCaffery, he distinguished between "art-fiction" that confronts readers and commercial art that pacifies them. The aim of the former was to "restore strange things' ineluctable strangeness, to defamiliarizes stuff", he said, thus recalling the Russian formalist view of the function of art to "de-automatize" perception (38). His critique of metafictional irony was not just that it had become coopted by the advertising industry, but that it had become mechanical: "after the pioneers always come the crank-turners, the little gray people who take the machines others have built and just turn the crank, and little pellets of metafiction come out of the other end" (ibid., 30). Fiction writers had turned into bureaucrats—"little gray people"—by no longer seeking to move readers into a state of heightened awareness but only seeking their approval. On the other hand, the effort to make fiction a form of self-confrontation in order to prevent it from slipping back into mere entertainment—"an anesthesia of form"—also entailed a risk (ibid., 32). The role of art for Wallace was to provoke thought, not block it, but the danger was that too much thought, especially in the form of self-consciousness, might itself become an obstacle to awareness. Zadie Smith (2009) puts the dilemma well when she writes that Wallace's "fear of automatism" led him in his work "to break the rhythm that excludes thinking", but also that "awareness must move always in an outward direction, away from the self" (264, 268). The unbroken rhythm of fiction creates an illusion that may become a palliative, but it also provides readers with an experience of aesthetic immersion that deepens concentration. Immersion may be lethal, as in the addictive entertainment provided by the eponymous film in *Infinite Jest*, but it may also be liberating in the sense of enhancing our power of attention. There is a fine line between distraction and attention, mindlessness and mindfulness. For Wallace, art should ideally provide the experience of immersion at the same time as it makes us aware of having that experience.

The episode in *The Pale King* narrated by Fogle dramatizes this when he recounts his experiences taking the drug Obetrol.[4] Describing the effect of the attention-enhancing drug as "doubling", he says that it gave him the ability to focus on his surroundings at the same time as he was conscious of focusing on them: "being in a state of heightened enough awareness to be able to consciously say to myself, 'I am in this room right now'" (181–182). He presents it as a de-automatizing experience—"As though I was a machine that suddenly realized it was a human being and didn't have to just go through the motions it was programmed to perform over and over"—yet one that nevertheless allows him to remain absorbed in the contemplation of his environment (182). Like Hawthorne in Sleepy Hollow, he is able to stay attentive to his surroundings while recording his impressions of them. Fogle explains that this drug-induced state of heightened awareness was one of "the factors that eventually drew me to the Service" (177).[5] Rote work may not offer the "doubling" experience Fogle is after, but it does hold the promise of immersion, which is the corrective he seeks to a culture in which the countercultural attempt to disrupt the "machine" of society has itself become mechanical. The IRS center in *The Pale King* has what it calls an "Immersives Room", where examiners work in "silent, motionless intensity" (290). It is precisely such a state of rapt attention enabled by mechanical work that the novel explores as a counterpoint to the "strange, robotic autopilot" of Fogle and his peers (185). "The underlying bureaucratic key", we are told, "is the ability, whether innate or conditioned, to find the other side of the rote, the picayune, the meaningless, the repetitive, the pointlessly complex. To be, in a word, unborable" (438). If pastoralism and deskwork used to be inimical, *The Pale King* relocates the pastoral garden from nature to the office. Alienation critique is turned on its head: the mechanical mind has become the new

4 Although the story does not mention this, Obetrol was not only a diet pill but in 1996 was reintroduced on the market under the new name Adderall and was prescribed for people diagnosed with A.D.H.D. See Alan Schwarz's (2013) article in *The New York Times* "The Selling of Attention Deficit Disorder".

5 The idea of a state of rapt attention was apparently also what drew Wallace to the IRS as a theme for his novel. D. T. Max (2012) quotes a letter Wallace received from an accountant on how the contemplation of a specific passage in the tax code could make him feel "a kind of weird Zen-type meditation high!", which appears to have inspired the connection *The Pale King* makes between rote work and meditation (291).

pastoral, the only way to counter the automatic mindset and instrumental behavior that the protests against automatism and instrumentality had led to.

Bye-Bye to the Bureaucrat

Yet true to the plot structure of "complex pastoralism", the prospect of bureaucratic repose in *The Pale King* is threatened by the invasion of a new force in society. The pastoral machine mind is imperiled by the arrival of a new machine in the landscape: the computer. If the garden of nature was destroyed by the appearance of the industrial machine, now the bureaucratic garden in the New Economy is being upset by the quintessential post-industrial machine. Although this plot element is little more than outlined in *The Pale King*, the intention is clear enough: "Big issue is human examiners or machines", we learn in the "Notes and Asides" (545). The narrative present of the novel is the mid-1980s, and bureaucracy is under attack on two fronts: ideologically from the neoliberalism of the Reagan administration; technologically from the personal computer. One character describes how the IRS in these years underwent a sweeping reorientation toward market values, and sums up the change thus: "As deregulated, there is now only one overriding operational question: Which returns are most profitable to audit, and how are those returns most efficiently to be found" (114). The answer to the second question is already given. "You are all, if you think about it, data processors", new IRS examiners are told (340). But if examiners are human data processors, then real data processors are surely more efficient. The efforts to reorganize the IRS on market principles are spearheaded by the novel's antagonist Dr. Merrill Errol Lehrl—a name evocative of the financial management firm Merrill Lynch—who aims to replace human examiners with computers. At the end of the novel, Lehrl has moved into the office of the civic-minded director of the regional IRS center, Mr. Glendenning, who appears to have been drugged and gone mad at a company picnic. One of the films discussed in *Infinite Jest* bears the title *Wave Bye-Bye to the Bureaucrat*. It deals with the "termination" of a bureaucrat, who chooses an act of compassion over saving his job (687–689). The IRS bureaucrats in *The Pale King*, however, have no such choice. Everything in the novel indicates that they will be run over by the neoliberal machine: a new era of marketized rules and computerized efficiency has dawned.

Alvin Toffler had already suggested as much in his 1970 book *Future Shock*: "Far from fastening the grip of bureaucracy on civilization more tightly than before, automation leads to its overthrow" (141). "As machines take over routine tasks", he predicted, "more and more of the energy of society (and its organizations) must turn toward the solution of non-routine problems" (ibid., 141). The picture he painted of society was not one of robotized workers and consumers, but one defined by the need to change and adapt to new rules and conditions at an ever-increasing pace. This was not a society of rigid bureaucratic hierarchies but of the accelerated thrust and permanent turnover of a "high-transience society" (ibid., 151). And in what he termed the new "Ad-hocracy", the psychological strain on workers did not result from understimulation or excessive discipline, but from overstimulation and the need for constant self-management in a world where nothing is fixed. The key for survival was not to be "unborable", but to be unshockable.

In its industrial nostalgia for the immersive possibilities of rote work in a low-transience milieu, *The Pale King* is clearly written against the backdrop of a world that looks more like Toffler's world of shock than Marcuse's world of repression. The minds of the most important characters in the novel are not mechanized but in a state of constant shock and inability to compose their thoughts. They do not seek self-fulfillment but self-abandonment, not to be awakened but calmed down, not freedom but order—or rather freedom through order. The solution is sought in the machine mind, but routinized tasks are being taken over by computers while job security is being undermined by a new entrepreneurial spirit in society. Ironically, the novel describes how the computerization of the workplace goes hand in hand with its ostensible humanization. One character explains how "the couple of years in question here saw one of the largest bureaucracies anywhere undergo a convulsion in which it tried to reconceive itself as a non- or even anti-bureaucracy", which he compares to "watching an enormous machine come to consciousness and start trying to think and feel like a real human" (80 fn). Another chapter consists of interviews with IRS employees as part of a publicity video for a campaign to "humanize, demystify the Service, help citizens understand how hard and important their job is. How much is at stake. That they're not hostile or machines" (100). Of course, machines are exactly what they will be replaced by. If the humanization of work used to be a countercultural issue, *The Pale King* shows how it has become a corporate ideology that functions to smooth the process of mechanization.

By satirizing this shift in corporate ideology, *The Pale King* is a timely neoliberal fable. It shows how the legitimacy of the New Economy relies on the critique of the old. By positioning bureaucracy as a middle landscape between the wilderness of the counterculture on one side and the neoliberal machine on the other, it dramatizes the ways in which oppositional culture and economy have converged in the new spirit of capitalism. As such, it would seem to confirm Leo Marx's skepticism about whether the pastoral surge on the New Left in the 1960s could form a viable political alternative, or whether, as a critique of a mechanical attitude in capitalist society rather than a critique of its economic framework, it might not be accommodated within the existing class structure. On the other hand, *The Pale King* does not pose the IRS bureaucracy as a solution to the problems of its characters because of what it stands for, but because of the psychological effect it has on them; not because it reduces inequality but because it calms them down. Inverted pastoral ideology is still pastoral ideology. While Giles pointed out how Wallace used "the human element to subvert rigid technocratic patterns", he did not note how the "human element" has itself become instrumentalized by corporate America (2007, 333). The consistent focus in Wallace's work, in his own words, on "what it means to be a fucking human", which critics have taken as a token of a new emotional sincerity in millennial fiction, is perhaps the best current expression of an alienation critique rebooted for the twenty-first century (1993, 26). The version of sincerity on display in work associated with the New Sincerity—the catchall term for art today that problematizes or seeks to transcend postmodern irony—does not imply the affirmation of values but the affirmation of feelings.[6] If it distances itself from the countercultural search for authenticity, it is because the counterculture has become inauthentic. But the pastoral impulse toward a simpler and more authentic life is unchanged. The difference is that whereas this impulse was radically at odds with how the economy was organized in the 1960s, today there is little friction between pastoral yearning and corporate ideology.

The emotional reflexivity at the heart of New Sincerity—summed up in the title of Nicoline Timmer's book on "post-postmodernism", *Do You Feel*

6 For the first reading of Wallace in light of the New Sincerity, see Kelly (2010).

It Too? (2010), which is a quote from Wallace's short story "Octet" (1999)—goes hand in hand with the growing market for skills that require emotional intelligence. "Unable to compete when it comes to calculation," David Brooks (2014) writes in a recent piece for *The New York Times*, "the best workers will come with heart in hand". *The Pale King*'s focus on attention as a scarce resource in a world where distraction is pathologized likewise feeds into the growing market for ways to stay focused. Commercial drugs for treating attention deficit disorder are marketed not only as enhancing the ability to focus but to make users more efficient.[7] Corporate self-help books today deal less with pop psychology—a genre which *The Pale King* parodies—than with technologies for self-micromanagement that train emotional skills as well as the capacity to stay calm and focused in a stressful environment (302). Books abound with titles such as *The Mindful Leader: Awakening Your Natural Management Skills Through Mindfulness Meditation* (2007), which features a green field with a tree on its front cover, or *The Power of Pause: How to be More Effective in a Demanding, 24/7 World* (2010). The endorsement of the latter by a CEO who promises that "this book is a profit accelerator and a buffer against stress", is suggestive of how new forms of attention management that help us take a "pause" in actual fact accelerate the pace of work, as they provide workers with techniques for handling ever more pressure. Numerous other examples of attention management could be mentioned, but it should already be clear that millennial pastoralism is not an alternative to the New Economy: it is a tool used to enhance it, and thus further corrode the old bureaucratic practices that *The Pale King* explores as a possible sanctuary for its characters. Assembled into a coherent whole by the book's editor Michael Pietsch, *The Pale King* ends on a note that echoes the pastoral mood with which it began. Only this time we are not in a pasture, but in a windowless office. As in the first chapter, the reader is addressed in the second person: "You are a trained observer and there is nothing to observe" (537). There are no distractions. And as in the opening, we are once more being guided by someone. The journey now, however, is not through a real landscape but a mental one. The person guiding us is identified as a "facilitator", and it becomes clear that she will guide us through a mindfulness meditation: "The way we start is to relax and become aware of the body" (538). On the one hand, the short final chapter is a culmination of the novel's exploration of deep at-

7 For Adderall's claims to enhanced performance, see Schwarz (2013).

tention and heightened awareness, just as the last sentence in the novel epitomizes the search for an emotional connection by reaching out to readers that characterizes Wallace's work: "She's right there, speaking calmly, and so are you" (538). On the other hand, the passage is also shot through with a streak of darkness that was missing in the vibrant description of an actual pastoral landscape in the first chapter. The lights have been dimmed "to the level of a false dawn", and the mental image that the meditation evokes of a staircase "curving down and rightward into a darkness that recedes before you" is anything but reassuring (538). In his 2005 Kenyon College commencement speech, which went viral on YouTube after his death, Wallace associated the power of attention and awareness with agency. "Learning how to think", he said, "really means learning how to exercise some control over how and what you think. It means being conscious and aware enough to choose what you pay attention to and to choose how you construct meaning from experience". This was Fogle's experience with awareness too: "It felt like I actually owned myself" (186). Yet by the end of *The Pale King*, this sense of control and ownership resulting from a mental attitude has become dubious at best. The pastoral mood in the last chapter does not suggest agency but a sense of defeat. The person addressed is completely passive, reclined in a chair and amazed at "what happens when someone else directs you how and when to breathe" (538). This is the novel's final vision of pastoral idyll, one that is managed by a company facilitator. As such, it questions the freedom that a pastoral mindset promises. If agency does not mean the power to change the social conditions that frame our lives, then the agency of a pastoral state of mind does indeed seem to augur a "false dawn".

Works Cited

Boltanski, Luc, and Eve Chiapello (2007 [1999]). *The New Spirit of Capitalism*. Trans. Gregory Elliott. New York: Verso.

Brooks, David (2014). What Machines Can't Do. *The New York Times*, Feb. 3. 11.02.2014 http://www.nytimes.com/2014/02/04/opinion/brooks-what-machines-cant-do.html.

Buell, Lawrence (1995). *The Environmental Imagination: Thoreau, Nature Writing, and the Formation of American Culture*. Cambridge: The Belknap Press of Harvard University Press.

Giles, Paul (2007). Sentimental Posthumanism: David Foster Wallace. *Twentieth-Century Literature*, 53.3, 327–344.

Guilmartin, Nance (2010). *The Power of Pause: How to be More Effective in a Demanding 24/7 World.* San Francisco: Jossey-Bass.

Hawthorne, Nathanial (1932). *The American Notebooks*. Ed. Randall Stewart. New Haven: Yale University Press.

Jameson, Fredric (1991). *Postmodernism, or, The Cultural Logic of Late Capitalism.* Durham: Duke University Press, 1997.

Kelly, Adam (2010). David Foster Wallace and the New Sincerity in American Fiction. In David Hering (ed.). *Consider David Foster Wallace: Critical Essays*, 131–146. Austin: SSMG Press.

Marcuse, Herbert (1970 [1955]). *Eros and Civilization*. London: Sphere Books.

Marx, Leo (1959). Notes on the Culture of the New Capitalism. *Monthly Review*, July–August, 111–116.

Marx, Leo (1967 [1964]). *The Machine in the Garden: Technology and the Pastoral Ideal in America.* New York: Oxford University Press.

— (1987 [1986]). Pastoralism in America. In Sacvan Bercovitch and Myra Jehlen (eds.). *Ideology and Classic American Literature*, 36–69. New York: Cambridge University Press.

— (1999/2000). Afterword. *The Massachusetts Review*, 40.4, 483–496.

Max, D. T. (2012). *Every Love Story is a Ghost Story: A Life of David Foster Wallace.* London: Granta.

Melville, Herman (1963 [1851]). *Moby-Dick*. New York: Signet.

Pietsch, Michael (2011). Everything and More: The Pale King by David Foster Wallace. Roundtable discussion at the PEN World Voices Festival of International Literature. 19.02.2014 http://www.pen.org/video/everything-and-more-pale-king-david-foster-wallace.

Schwarz, Alan (2013). The Selling of Attention Deficit Disorder. *The New York Times*, Dec. 14. 19.02.2014 http://www.nytimes.com/2013/12/15/health/the-selling-of-attention-deficit-disorder.html.

Smith, Henry Nash (1981 [1950]). *Virgin Land: The American West as Symbol and Myth.* Cambridge, MA: Harvard University Press.

Smith, Zadie (2009). Brief Interviews with Hideous Men: The Difficult Gifts of David Foster Wallace. In *Changing My Mind: Occasional Essays*, 257–300. London: Penguin.

Toffler, Alvin (1990 [1970]). *Future Shock*. New York: Bantam.

Wallace, David Foster (1989/2011). Westward the Course of Empire Takes Its Way. In *Girl with Curious Hair*, 231–373. London: Abacus.

— (1993/2012). Interview with Larry McCaffery. In Stephen J. Burn (ed.). *Conversations with David Foster Wallace*, 21–52. Jackson: University of Mississippi Press.

— (1996). *Infinite Jest*. New York: Little, Brown and Company.

— (2004/2012). Interview with Steve Paulson. In Stephen J. Burn (ed.). *Conversations with David Foster Wallace*, 127–135.

— (2005). Kenyon College Commencement Address (Transcript). 19.02.2014 http://moreintelligentlife.com/story/david-foster-wallace-in-his-own-words.

Wallace, David Foster (2011). *The Pale King*. New York: Little, Brown and Company.

Wood, James (2000). Human, All Too Inhuman. *The New Republic*, July 24. 10.02.2014 http://www.newrepublic.com/article/books-and-arts/human-all-too-inhuman.

Epilogue: Politics and Culture in *The Machine in the Garden*

Alan Trachtenberg

In 1958 I took a graduate seminar with Leo Marx at the University of Minnesota. The seminar was in many regards a preview of Marx's major book of 1964, his formative study of "machine" and "garden" and the fraught preposition "in." The class focused on interlocked meanings of key words like these, including industrialization, industrialism, and pastoralism, terms charged with powerful conceptual and emotive meaning within the evolving academic field of American Studies. A half a dozen years before the actual book appeared, the seminar inducted us, so to speak, into the interrelated discourses that formed the basis in cultural history of the new field. Before the hard copy appeared, as it were, we were initiated into the discourse of "the machine in the garden." Marx's charismatic teaching made us eager to embrace and put into practice this new line of thought in the academic study of "the American view of life," a phrase from the book which appeared as the title of a review of the book I published in *The Nation* in July 1965.

Reviewing the book for a major journal of politics and arts, a periodical with an unmistakable progressive or left perspective, gave me an opportunity to think about the book in light of its method as well as its argument, its form, style, and structure as a politically-engaged example of a relatively new mode of literary scholarship. "American Studies" had emerged in the 1930s and 1940s as an effort to integrate separate aspects of culture into a holistic interdisciplinary vision of a whole social entity. I was then in my first job after graduate school, an assistant professorship of English at a large state university. I had come to the Minnesota program in American Studies from an almost exclusively literary background yet with a strong bent toward historical and social—a generally Marxist bent—approaches to literary study. The dazzling literary interpretations in Leo Marx's book immediately turned my head and filled me with delight and desire to follow suit. Close textual readings of literary responses to industrialization in *Wal-*

den, *Moby Dick*, *Huck Finn*—in many ways suggestive of the ahistorical readings practiced by the so-called "New Critics" in the same years—are embedded in a flow of countless eye-opening insights into works by Ralph Waldo Emerson, Walt Whitman, Henry James, F. Scott Fitzgerald, and others. Sharply focused on specific texts and passages, the book is informed by a continuous stream of sensitivity to literary language and form.

But what spoke most directly to the need I felt as an aspiring scholar of American Studies and has remained the major bequest of the book for me, even beyond the brilliant originality of its dialectical opposition of industrial-urban and pastoral, was and remains the book's determined historicism. I mean its particular kind of historicism that Marx himself, in his eloquent "Afterword," names as "historical materialism" (Marx 2000, 484), a term that evokes another figure named Marx—Karl Marx—who figures in the book in ways worth further study. The role of Marxism in Leo Marx's work, palpable even if not addressed explicitly in the book, was doubtless filtered through his personal relationship with prominent America socialists, the economist and author Paul Sweezy, co-editor with Leo Huberman of the independent socialist journal *Monthly Review*, which had been founded in 1948 with funds made available by another of Marx's important friends and teachers, F. O. Matthiessen, Harvard professor of English and American Literature and socialist critic of American society and politics in the opening years of the Cold War.

Yet the relation of the book to Marxism and socialism has hardly been prominent or even at all evident in the book's reception and continuing influence. Fuller appreciation of the book's argument and its character as a work of literary-historical-cultural writing awaits deepening with a restored awareness of this dimension of *The Machine in the Garden*—its dimension as a pointed critical response to a menacing historical situation such as it prevailed in post-war America and the opening years of the partisan divisions and conflicts of the Cold War. The book showed that a certain kind of literary-cultural discourse could bring together diverse materials—historical, cultural, literary, visual, political—into a subtle and complex response to its own moment of production. That moment of production or the book's setting during the early phase of Cold War America becomes even sharper, more lucid and comprehensible, when confronted by the array of critical materials the book evokes and places in relation to it. The book understands that history cannot be reduced to an array of facts but that it takes shape as a set of relationships and cross-currents: not the railroad engine alone so starkly depicted as a black form against a field of green on the book's original cover, but in a relation with the "pastoral ideal" as the conceptual link or mental connective tissue provided by the literary and political culture. "Machine in the Garden" forms the field upon which the dialectical drama at the heart of the book occurs, real space melded with imagined and imaginary space.

We have to add, too, that we cannot speak of the politics of the book, the vision of American actuality and possibility that Marx plants in the nervous system and at the heart of his discourse, without recognizing that the story Marx tells of the fate of the pastoral ideal in the setting of the conflict between machine and garden has no happy outcome. Refracted in the imagination of certain writers and artists, the conflict gave rise to major artistic works which have, Marx avers, clarified "the root conflict of our culture" (Marx 2000, 365). But conceptual clarification alone does not alter the basic predicament or array of social forces. The machine remains dominant over the social order and its nostalgic echoes and hints of Jeffersonian democracy. The book tells one story after another of failure, failure of hope, of vision, of imagination of viable alternatives to the capitalist-industrial order in which the machine reigns. The great texts in which this failure of pastoral ideals—their failure measured as the failure of viable alternatives to the existing social order—"have served us well," Marx concludes in his final sentences, which reaffirm the resolute historicism of his approach, his effort to hold history and art together in a single vision: "To change the situation we require new symbols of possibility, and although the creation of those symbols is in some measure the responsibility of artists, it is in greater measure the responsibility of society. The machine's sudden entrance into the garden presents a problem that ultimately belongs not to art but to politics" (ibid., 365).

This extraordinary work of 1964 concludes then, with a call that still rings with relevance. Perhaps not enough readers of the book have paid enough attention to its implied (yet actively assumed) political commitment, its commitment to a progressive politics of democratic social change. In the book's assumption that the political serves as the necessary ground and condition of what we value in literature and culture, not an after-thought or a supplement but a dimension of experience intrinsic to all our concerns, *The Machine in the Garden* remains a work to live up to.

Works Cited

Marx, Leo (2000 [1964]). *The Machine in the Garden: Technology and the Pastoral Ideal in America.* London: Oxford University Press.

— (1999). Afterword: The Machine in the Garden. *The Massachusetts Review*, 40.4, 483–496.

Trachtenberg, Alan (1965). The American View of Life. *The Nation*, 201, 42–45.

Notes on Contributors

Laura Bieger is Professor for North American Literature and Culture at Albrecht-Ludwigs-Universität Freiburg. She is the author of *Ästhetik der Immersion. Raum-Erleben zwischen Welt und Bild* (Transcript, 2007). Forthcoming and recent publications include: *No Place Like Home. A Narrative Theory of Belonging* (forthcoming 2015), "Space, Place and Narrative," Special Issue of *Zeitschrift für Anglistik und Amerikanistik*, ed. with Nicole Maruo-Schröder (forthcoming 2015), *The Imaginary and its Worlds: American Studies after the Transnational Turn*, ed. with Ramón Saldívar and Johannes Voelz (University of New England Press, 2013), and *Revisiting the Sixties: Interdisciplinary Perspectives on America's Longest Decade*, ed. with Christian Lammert (Campus, 2013).

James Dorson currently has a DFG Grant "Eigene Stelle" at the John F. Kennedy Institute of North American Studies, Freie Universität Berlin for his research project "The Aesthetics of Mastery: American Literary Naturalism and the Rise of Managerialism." In 2011 he has completed his doctoral degree with a dissertation on the notion of counternarratives in Cormac McCarthy's novels. He has several articles on Cormac McCarthy, David Foster Wallace, and Edith Wharton, published with peer-review journals and in essay collections.

Eric C. Erbacher is a Lecturer in American Studies at the Westfälische Wilhelms-Universität Münster. His research interests are urban studies, media studies, and cultural studies, with a particular focus on the intersection of these fields with issues of visual culture, economic restructuring, popular culture, and class. He is currently working on a project on the narrative construction of space in American gentrification processes. He has published articles on social, ethnic, and economic issues in urban transformations, on contemporary popular (sub-)cultures, and on TV series. He is also the co-founder of the SpaceCollective, an interdisciplinary network of young scholars in the Ruhr Area interested in spatial theories and practices.

Christa Grewe-Volpp is Professor of American Literature and Culture at Universität Mannheim. Her research focuses on ecocriticism and ecofeminism, on ethnicity and postcolonialism, and on issues of gender. She is the author of a book on ecocritical and ecofeminist analyses of contemporary American writers: *"Natural Spaces Mapped by Human Minds": Ökokritische und ökofeministische Analysen zeitgenössischer amerikanischer Romane* (Narr, 2004). She also co-edited (with Maureen Devine) an ecocritical collection of essays: *Words on Water: Literary and Cultural Representations* (Wissenschaftlicher Verlag Trier, 2008), and she edited a thematic section of the ecocritical journal *ecozon@*: "Greening Across Borders: the Natural Environment in a Globalized World" (2010).

Susanne Hamscha teaches American Studies at Universität Göttingen. Currently, she is a Visiting Scholar at Columbia University, where she is working on her postdoctoral project on the American freak show tradition. Susanne completed her doctoral degree in North American Studies at Freie Universität Berlin in 2012. Her dissertation *The Fiction of America: Performance and the Cultural Imaginary in Literature and Film* was published with Campus in 2013. She is also co-editor of the volume *Is It 'Cause It's Cool? Affective Encounters with American Culture* (LIT-Verlag, 2014) and the forthcoming *Freakery: Cultural Spectacles of the Extraordinary Body, Volume 2* (New York University Press).

Katja Kanzler is Professor of North American Literature at TU Dresden. Her work focuses on the intersectionalities of race, class, and gender in U.S. American literature and popular culture, on genres of popular culture past and present, and on the dynamics of narrativity and textuality in different genres and media. Her article in this book is part of a larger project that explores how and to what cultural ends antebellum literature represents spaces of women's work.

Marlon Lieber is research and teaching assistant at the Institute for English and American Studies at Goethe-Universität Frankfurt am Main. He is currently working on his doctoral dissertation on the novels of Colson Whitehead tentatively entitled "'You're Not Going to Tell Me How it

Turns Out?' Colson Whitehead's Novels and the Ends of 'Race.'" In it he is using Pierre Bourdieu's sociology and particularly the theory of symbolic violence in order to analyze the role race plays in the novels of this author that has been referred to as "post-racial" or "post-black." In addition to relational sociology and African American literature his research interests include critical theory and the critique of political economy.

Nicole Maruo-Schröder is Professor of Cultural Studies at Universität Koblenz-Landau. Her teaching and research interests include space, place, and narrative; contemporary American film; ecocriticism as well as the interconnections between material culture and literature. Currently, she is working on a project that explores the portrayal of consumption and consumer practices in antebellum American literature. Recent and forthcoming publications include a volume on *Literature and Consumption in Nineteenth-Century America* (Winter, 2014), edited with Christoph Ribbat, and "Space, Place and Narrative," Special Issue of *Zeitschrift für Anglistik und Amerikanistik*, edited with Laura Bieger (forthcoming 2015).

Dietmar Meinel was a doctoral fellow at the Graduate School of North American Studies at the John F. Kennedy Institute, Freie Universität Berlin from 2010 to 2013. His dissertation project explores animated film, U.S. exceptionalism and aesthetic theory under the working title "Exceptional Aesthetics. Reconsidering Aesthetic Experience in the (New) American Exceptionalism of Pixar Films" supervised by Winfried Fluck, Laura Bieger, and Donald Pease. Since October 2013 he is a research assistant in American Studies at Universität Duisburg-Essen. His research interests focus on mainstream as well as independent film, contemporary popular culture, and visual arts particularly within the context of American Studies In 2008, he has edited the book *Black, White & In-Between* together with Laura Bieger and contributed the essay "White Western. Whiteness and Race Politics from John Wayne to Clint Eastwood" to the collection.

Florian Sedlmeier is Junior Professor of North American Literature at the John F. Kennedy Institute, Freie Universität Berlin. His monograph *The Postethnic Literary: Reading Paratexts and Transpositions around 2000*, which is based upon his doctoral thesis submitted at Universität Salzburg in 2010, will be published with de Gruyter in 2014. Two of his essays have been published with the *Journal of Literary Theory* and articles on Ralph Ellison's unfinished and posthumously published second book, notions of cover and transparency in the television format *Glee*, and the iPad are included in scholarly volumes. Essays on the relation of absorption and theatricality in Gertrude Stein's "Plays" and on the poetics of seriality in Don DeLillo's *Falling Man* are forthcoming. The volume *Kriminalliteratur und Wissensgeschichte*, which he co-edits with Clemens Peck, is scheduled for publication in spring 2015.

Klara Stephanie Szlezák received her doctoral degree in American Studies from Universität Regensburg in 2013. Currently, she prepares her dissertation *'Canonized in History': Literary Tourism and Nineteenth-Century Writers' Houses in New England* for publication and works on an edited volume on Woody Allen. She has published articles on literary tourism and on early twentieth-century photography and immigration, and is a co-editor of the e-journal *COPAS—Current Objectives in Postgraduate American Studies.*

Alan Trachtenberg is Neil Gray, Jr. Professor Emeritus of English and American Studies at Yale University. A former student of Leo Marx's, he is the author of *Brooklyn Bridge: Fact and Symbol* (University of Chicago Press, 1965), *The Incorporation of America: Culture and Society in the Gilded Age* (Hill and Wang, 1982), *Reading American Photographs: Images as History, Matthew Brady to Walker Evans* (Hill and Wang, 1990), *Shades of Hiawatha: Staging Indians, Making Americans, 1880–1930* (Hill and Wang, 2004), and the essay collection *Lincoln's Smile and Other Enigmas* (Hill and Wang, 2007). He also edited the volume *Classic Essays in Photography* (Leets Island Books, 1981).

Kirsten Twelbeck has just completed a long-term research project ("Habilitation") at Leibniz-Universität Hannover. In this project, which is to be published as a book, she investigates how white American women and men imagined the future nation during the American Civil War and Reconstruction. She holds a doctoral degree in American Studies from Freie Universität Berlin in Berlin. Her dissertation is a study on the representation of Korean Americans since 1965; it has been published under the title *No Korean Is Whole – Wherever He Or She May Be* (Lang, 2001). Kirsten Twelbeck currently works as an Assistant Professor of American Literature and Culture at Leibniz-Universität Hannover. Her general research and teaching interests include the historical relationship between the United States and Asia, postcolonial studies, gender studies, medical history, photography, and questions of nation building and citizenship in nineteenth- and twentieth-century American culture. Moving from contemporary debates in the field of Asian American Studies to the broader historical development of intercultural interactions, she has worked on women in the Christian missionary movement in the United States, on South African literature, and on the representation of Europe in contemporary Jewish American literature.

Index